P9-BYR-010

MACHINING
FUNDAMENTALS

From Basic to Advanced Techniques

JOHN R. WALKER

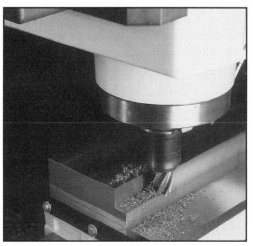

Publisher
The Goodheart-Willcox Company, Inc.
Tinley Park, Illinois
www.g-w.com

The Goodheart-Willcox Company, Inc. Brand Disclaimer: Brand names, company names, and illustrations for products and services included in this text are provided for educational purposes only and do not represent or imply endorsement or recommendation by the author or the publisher.

The Goodheart-Willcox Company, Inc. Safety Notice: The reader is expressly advised to carefully read, understand, and apply all safety precautions and warnings described in this book or that might also be indicated in undertaking the activities and exercises described herein to minimize risk of personal injury or injury to others. Common sense and good judgment should also be exercised and applied to help avoid all potential hazards. The reader should always refer to the appropriate manufacturer's technical information, directions, and recommendations; then proceed with care to follow specific equipment operating instructions. The reader should understand these notices and cautions are not exhaustive.

The publisher makes no warranty or representation whatsoever, either expressed or implied, including but not limited to equipment, procedures, and applications described or referred to herein, their quality, performance, merchantability, or fitness for a particular purpose. The publisher assumes no responsibility for any changes, errors, or omissions in this book. The publisher specifically disclaims any liability whatsoever, including any direct, indirect, incidental, consequential, special, or exemplary damages resulting, in whole or in part, from the reader's use or reliance upon the information, instructions, procedures, warnings, cautions, applications, or other matter contained in this book. The publisher assumes no responsibility for the activities of the reader.

Cover photos courtesy of Sandvik Coromant Co., Iscar Metals, Inc., and O. S. Walker Co.

Contents

Introduction

1.0 OVERVIEW

Machinists are highly skilled men and women. They use drawings, hand tools, and various machine tools to shape and finish metal and nonmetal parts. Machinists must have a sound understanding of basic and advanced machining technology, including:

- The ability to properly use precision measuring tools.
- A proficiency in safe machine tool operation (manual, automatic, and computer-controlled).
- A knowledge of the working properties of metals and nonmetals.
- Basic academic skills (math, science, English, print reading, metallurgy).

A well-planned machining program integrates and balances hands-on experience with the comprehensive coverage of the technical aspects of machining technology.

Machining Fundamentals provides an introduction to this important segment of manufacturing technology. It furnishes basic information on the tools, materials, and procedures employed in machining technology. Remember—*before* you can go high tech, you must first understand the fundamentals.

1.1 INSTRUCTIONAL MATERIALS

The following materials have been developed to aid in presenting a dynamic machining program.

1.1.1 Textbook

The textbook is a very important part of an instructional program. *Machining Fundamentals* provides an introduction to machining technology. The text explains the *how, why,* and *when* of the various machining operations, setups, and procedures. Through it, your students/trainees will learn about the various related areas of machining technology, how machine tools operate, and when to use one particular machine instead of another. The advantages and disadvantages of the various machining techniques are also discussed.

Machining Fundamentals details the many common methods of machining and shaping parts to given specifications. It also covers newer processes such as laser machining and welding, water-jet cutting, high energy rate forming (HERF), cryogenics, chipless machining, electrical discharge machining (EDM), electrical discharge wire cutting (EDWC), electrochemical machining (ECM), numerical control (NC and CNC), robotics, rapid prototype forming, and the importance of computers in the operation of most of these machining techniques.

Machining Fundamentals is written in an easy-to-understand language. There are many color photographs and line drawings to help students/trainees clearly visualize machining operations and procedures.

Colors are used throughout *Machining Fundamentals* to indicate various materials or equipment features. The color key is on page 4 of the text. The following list identifies colors and materials by name.

Metals (surfaces)	Dark Gray
Direction or force arrows, dimensional information	Red
Machines/machine parts	Blue

Fasteners   Yellow

Tools   Dark Green

Abrasives   Brown

Work-holding and
  tool-holding devices   Purple

Fluids   Lime Green

Rulers and
  measuring devices   Tan

Miscellaneous   Orange

Learning Objectives

Every chapter opens with a list of objectives. They make the student/trainee aware of what he/she will be able to accomplish after studying the chapter.

Because of many factors, it may not be possible to achieve every objective with each class. However, it is better to thoroughly cover less material than to cover all of the material poorly.

Test Your Knowledge Questions

Each chapter ends with a set of questions. The questions can be used to check student comprehension of the text material. The questions can be used as either a quiz, homework, or extra credit.

Glossary

The glossary provides a quick reference to the definitions of technical terms used in machining technology.

Reference Section

A reference section of tables and other useful information is provided on pages 557–592. This section contains information on a variety of topics including the physical properties of many metals, cutting speeds and feeds, drill and screw thread sizes, and metric tables.

1.1.2 Instructor's Resources

The resources have been devised to assist the instructor in improving their machining technology program. For each chapter, it includes *Learning Objectives*, a list of *Technical Terms*, the text's *Test Your Knowledge Questions*, *Research and Development Ideas*, *Answer Keys*, *Reproducible Masters*, and *color transparencies*. It also includes a correlation chart for Level I of the *Duties and Standards for Machining Skills*.

Learning Objectives

For ease in referencing, text objectives are listed for each chapter. The goals presented involve basic concepts, skills, and understand- ings that should be stressed while teaching the chapter. The objectives can also be used as an outline when preparing lesson plans.

Technical Terms

It is important that machinists use the correct technical terms of the trade. Students should be encouraged to review and study any new and unfamiliar terms.

Test Your Knowledge Questions

The questions from the end of the chapters are included as reproducible masters. Save time by copying and distributing the reproducible master rather than having students use the questions at the end of the chapter and writing their answers on a separate sheet of paper.

Research and Development Ideas

These ideas offer an opportunity to bring many of the latest machining techniques into the shop/lab. They can be used by both the instructor and the student.

They can be used for open discussion, homework assignments, extra credit, or individual or group projects. They provide the opportunity for developing student/trainee originality and ingenuity.

Answer Keys

Answers to the *Test Your Knowledge* questions and workbook are provided within each chapter.

Reproducible Masters

Full-page illustrations and *Test Your Knowledge* questions that can be copied and distributed to students and/or used as overhead transparencies. Each is correlated to the text with a chapter identification number. Color can be added to the transparencies with felt tip pens.

Color Transparencies

A set of full-color transparencies has been produced to help the instructor reinforce concepts presented in the textbook. The 56 transparencies are available in printed form with the Instructor's Resource Binder and in electronic form with the Instructor's Resource CD.

Duties and Standards for Machining Skills Level I *Correlation Chart*

A chart correlating the *Duties and Standards for Machining Skills* to the **Machining Fundamentals** text has been included on pages 25–26 of this resource. The categories are correlated to the text by chapter and section numbers.

The standards were developed by the metal-working industry to provide a certification and training method through which individual workers can receive recognition and reward for their abilities. The standards will also help employers identify training needs and evaluate job applicants fairly.

Although elements of all three standards often appear in training programs of varying levels, *Machining Fundamentals* is an introductory text so only Level I of the standards will be covered in this manual.

For additional information on the standards and testing programs, contact the National Tooling and Machining Association (NTMA), 9300 Livingston Road, Fort Washington, MD 20744.

G-W Test Creation Software

The G-W Test Creation Software is included with the Instructor's Resource CD. The database for the software package includes over 1400 questions. The versatility of the software allows the instructor to create customized tests.

Tests can be generated with randomly selected questions, the selection of specific questions from the database, and the addition of personalized questions. Different versions of the same test can be created for use during different class periods. The software also allows the modification and importing of graphics. Answer keys are automatically generated to simplify grading.

1.1.3 Workbook

The workbook is an aid for measuring student/trainee achievement and comprehension. It uses a variety of questions, problems, and assignments. The workbook materials within each chapter are presented in the same order as corresponding material in the textbook.

For use as a study guide, students/trainees should first read and study the material assigned in the textbook, giving careful consideration to the illustrations. Then, without the aid of the textbook, fill in the workbook answers. As recommended for several chapters of the text, the assignments in the workbook can also be divided into parts as seen fit by the instructor.

Answers to the questions and problems consist of words, letters, numbers, and simple drawings. Instruct students that words should be spelled correctly and letters and numbers should be carefully formed. It is highly recommended that the letters and words be printed. Stress to students that most tradespeople follow this same practice since the information will be easier to read and the possibility of errors greatly reduced. Sketches should be carefully drawn in the space provided. When required, mathematical calculations should be made in a neat, organized manner. This makes it easier to check the procedure used in solving the problem.

1.2 TO THE INSTRUCTOR

It is not possible for an author or publisher to provide a detailed machining technology program that will be suitable for every teaching situation. Some training programs require a specific outline to be followed. Also, the type, number, age, and condition of the machine tools available will vary.

As the instructor, only you can determine the material that will best serve your students/trainees. Only you know their abilities, the facilities, materials, and equipment at your disposal, and the time available.

Before the first class meeting, familiarize yourself with the text and its related teaching material. Outline the chapters you plan to teach. Prepare detailed lesson plans and gather and/or make the teaching aids that will be needed. Check equipment that will be used. Be sure it is in good condition with all safety features in place. Have ample supplies on hand.

This will probably be the first machining technology class to which your students/trainees have been exposed. What is so obvious to you, may be completely foreign to them. Make your lesson plans with this in mind.

1.3 SHOP/LAB MANAGEMENT

There are several areas in shop/lab management that must be considered. If properly developed, they can save time that can better be devoted to one-on-one teaching.

1.3.1 Control of Tools and Consumable Supplies

Many expensive tools are required in machining technology. A method for controlling their disbursement and return to inventory must be devised and continually monitored to reduce the number of damaged tools and to prevent pilferage. The same care must be exercised for issuing stock and other consumable supplies.

You can examine the systems local industries have established for control of tools and supplies, and implement one of them, or you can devise your own.

1.3.2 Scheduling

Because of equipment limitations, you will not be able to have all students/trainees working on the same assignment at the same time. Assignments must be organized and scheduled so equipment will be used as much as practical.

1.3.3 Shop/Lab Management System

Good housekeeping and cleanliness is important in machining technology. Insist that students/trainees clean workstations after use. To aid in overall shop/lab cleanliness, each person should be assigned, on a rotating basis, a specific cleanup task each week. Praise them when the job is well done.

Once a week, a more thorough cleanup should be done. This includes cleaning lockers and getting shop coats and aprons washed.

Since the best teaching is on a one-to-one basis, a well organized student/trainee management system will give you more time for individual aid.

Assignments should be given on a rotating basis so all of them will be able to experience each position at least once during the term.

In the interest of maintaining a safe and orderly shop, a student/trainee management system should include the following positions and responsibilities:

Superintendent
Responsibilities should include:
- Directing the student/trainee management system.
- Bringing the class to order.
- Inspecting the shop/lab for cleanliness and apparent or potential safety problems.
- Checking with other systems managers to be sure they are working at their assigned duties.
- Assigning students to fill in system vacancies caused by absences.
- Initiating cleanup.

- Inspecting the shop/lab at the end of the period for cleanliness, securing reports from other systems managers.
- Reporting the condition of the machines, tools, and supplies.

Records Clerk
Responsibilities should include:
- Recording attendance and tardiness.
- Keeping progress charts up-to-date.
- Serving as bookkeeper for stock requisition.
- Delivering messages.
- Greeting visitors.

Tool Crib Manager
Responsibilities should include:
- Issuing and returning tools to proper storage.
- Keeping accurate records of tools issued and returned.
- Reporting missing tools and tools in poor or unsafe condition.

Stock and Supply Supervisor
Responsibilities should include:
- Recording student/trainee requisitions for metal stock and supplies.
- Cutting stock to requisitioned size.
- Sorting metal scrap to salvage usable material.
- Reporting pending shortages of supplies.
- Keeping stock organized in storage racks.

The above are only suggested management positions. They can be added to, and the duties modified and/or changed.

1.4 IMPROVING INSTRUCTION

Whether teaching machining technology, or any other subject, you make use of certain universal instructional tools. All good teachers apply these concepts, either consciously or unconsciously. When making your lesson plans, try to implement the ideas listed below.
- *Reinforce.* The more ways a student/trainee is exposed to a given concept, the greater the understanding and retention of the material. A variety of learning experiences are designed to meet the reinforcement needs of the learner.
- *Extend.* The teaching suggestions in this manual are directed at students/trainees with a variety of ability levels. Some assignments may be chosen to encourage highly motivated students/trainees to extend their learning experience outside the shop/lab. These types of activities allow them to relate text information to other experiences.

- *Reteach.* Students/trainees respond differently to diverse teaching methods and techniques. This allows you to choose a different plan of action to reteach those who responded poorly to a previous strategy.

1.5 PLANNING AN INSTRUCTIONAL PROGRAM

No matter what approach you take in teaching a basic machining technology program, the importance of careful planning and organizing *cannot* be overemphasized. Planning to achieve program goals will be easier if you know *why, when,* and *how* to plan.

Knowing *Why* to Plan

Planning is the process of carefully selecting and developing the best course of action to achieve program goals and objectives. This action cannot be a hit-or-miss situation. How well you meet your teaching responsibility depends, to a great extent, upon your consistency. Every class should be planned and prepared carefully.

We plan so action will take place at the right time. Planning helps anticipate problem areas of learning, and makes adapting to and handling emergencies easier. Planning saves time, money, and ensures a higher quality product—your students/trainees.

Planning helps assure quality results, saves valuable class time, helps reduce discipline problems and makes it easier to adapt to changes. Planning is also good teaching— knowing what has to be taught, what has been taught, and what needs to be taught to reach program goals and objectives.

Knowing When to Plan

In lesson planning, it is best to be flexible. That is, planning must be continuous. You must be able to adjust your lessons for differences in classes or classes missed because of bad weather and other unplanned interruptions.

Rigid planning means that a specific lesson will be taught on a specific day. If a class is missed that day, there is no way to make it up. Although preferable to not planning at all, rigid planning is less effective than flexible planning.

When developing plans, you must consider the time available. Once you know this, you can classify the demands on this time by what must be done, what should be done, and what need not be done.

After a course of action has been determined, specific planning can begin by:

- Establishing goals for each class. Some classes may require more time to master the same skills and information than other classes.
- Taking inventory of supplies and equipment. Be sure the supplies are adequate.
- Considering different teaching techniques. What other methods are available that will be equally effective?
- Providing specific learning experiences in an interesting manner.
- Setting plans into action. Once a plan of action has been developed, it should be followed.

Knowing How to Plan

A lesson plan is an outline for teaching. It keeps the essential points of the lesson in front of you and ensures an orderly presentation of material. Such a plan does much to prevent important aspects of a lesson from being omitted. It will also prevent you from straying too far from the lesson and introducing irrelevant material.

Your lesson plan should include the material to be taught, the methods and techniques best suited to teach this material, any supplies or equipment needed, and teaching aids to be used.

Review your material before class time. Be familiar with all of the material related to the subject. You should be prepared to teach more material than you may have time to cover. Check all of the supplies and equipment to be sure they are ready to use, in safe working condition, and arranged so they will be handy during the lesson. Seating and viewing arrangements should be such that every student/trainee can see and hear the lesson.

Finally, keep your plans up-to-date. This can be done by observing student/trainee progress and achievement. Make changes when you think they are necessary for curriculum improvement.

In general, *always* prepare a lesson plan and *follow it.*

Modify it as necessary. Start with something that is familiar to the student/trainee. Then move into new material in short easily understood steps. Avoid long boring lessons and lessons that do not allow student participation.

The First Class Session

After becoming familiar with *Machining Fundamentals* you will have to determine which chapters can be used in your machining technology program and to the extent to which they will be covered. This will require special planning so students/trainees will acquire sufficient skills development through "hands on" activities.

Whatever your selection, Chapters 2, 3, and 4 should *not* be omitted. They are basic to any machining technology program.

During the first class session, after all the administrative paperwork is completed, have your students/trainees complete a personal information sheet designed like a job application. In addition to giving them experience in filling out a job application, it will help you get to know them better.

1.6 TEACHING METHODS FOR MACHINING TECHNOLOGY

No learning takes place until a student/trainee wants to learn. Getting them interested is the *most* difficult part of teaching. Motivation, therefore, is the first step in good teaching. You want to stimulate students/trainees so they want to learn.

One of the quickest way to lose student interest is to be unprepared and present a meandering lesson. For effective teaching, you must carefully prepare the material, the situation in which it will be taught, and the student/trainee who will receive the new information.

Start by learning and mastering the material yourself. Be sure all necessary materials and teaching aids are readily available and in good working condition.

Determine what equipment will be needed and position it so all students can see and hear the lesson and/or demonstration.

Decide how you want to motivate your class. You can use curiosity. Students/trainees want to see, hear, and know about new and different things. Competition is another way to create interest. Some learners may want to take on a challenge to surpass another person or group.

Above all, you must be interested in what you are doing. If you are not interested in the subject matter, it is unlikely your students/trainees will be interested.

Before placing responsibility for failure on a student, review and evaluate the lessons taught to determine whether the goals of the lessons were attained.

Some instructors assume that if they are quick to learn, the same holds true of others. This has the effect of retarding the learning process rather than enhancing it. Impatience can create student/trainee fear of the instructor. To do a good job, you must have the respect of the learner. You do not have to be liked by them, but you do not want to be feared. There is quite a difference.

Although careful preparation will greatly improve your teaching effectiveness, it does not guarantee success. The amount of effort to develop and hold student interest will vary from class to class. More or less instruction than was planned may be required. Student/trainee alertness may vary with physical condition including illness, fatigue, and lack of sleep. Their rate of learning may change for no apparent reason. There may be distractions that you did not expect.

In order to determine what may be distracting your students/trainees, or causing lack of interest, consider the following. Do you:

- Use and pronounce words correctly?
- Avoid the use of localisms, slang, and monotonous connectives?
- Look directly *at* and speak directly *to* the learner?
- Maintain good eye contact?
- Use the appropriate vocabulary level?
- Prepare yourself?
- Use variations in the pitch of your voice?
- Give credit where credit is due?
- Stimulate thinking when you ask a question by phrasing it to bring out the *why* and *how*?
- Know all students by name?
- Summarize frequently, as each major point in the lesson is made?
- Direct questions at inattentive students/trainees?
- Delay the entire class when one student is causing a problem?
- Continually check class reaction?
- Use teaching aids whenever possible?
- Display teaching aids where the entire class can see them?
- Check training aids like video equipment, projectors, screens, and tapes before class starts?
- Preview and select audio-visual material for specific instructional content?

- Prepare students/trainees for the video, film, or tape?

The list could go on and on but the key is being prepared. Know what you are going to teach and how you are going to do it. You can only develop student interest in what you are teaching if you are interested in what he/she is achieving.

Some other tips to teaching effectively include those listed below.

- Put the students at ease. Humor is an effective tool to help students/trainees to relax. Remember, a tense atmosphere will inhibit learning.
- Students have five senses. The more senses you can involve in the learning process, the more likely the student/trainee will remember what is taught.
- You should direct questions to individual student/trainees to attract and hold attention and interest, measure knowledge and understanding, and focus attention on the main concepts.
- Students/trainees are people with different personalities and attitudes. You will be more successful as a teacher if you get to know your students individually. This will allow you to see the areas in which each one needs help and encouragement.
- Do not allow the class to become a monologue in which you spend the entire period lecturing. You must encourage them to ask questions and discuss topics.
- When using a chalkboard, stand to one side of the board. Also, talk directly to the class *not* to the chalkboard.
- Remember, topics which are simple to you may be most complex for the students/trainees. Avoid covering a topic too quickly.
- You must speak clear enough, loud enough, and slow enough for your students/trainees to follow the lesson.
- Use a tape recorder to critique your verbal presentation.

1.6.1 Student/Trainee Evaluation

Discuss with the students/trainees how they will be evaluated. Emphasize the importance of attendance and its significance when they enter the workplace. Avoid allowing students/trainees with a number of unexcused absences to make up missed work at the end of the work term with a short meaningless assignment. Some programs require a predetermined number of hours in actual attendance before a certificate or degree is granted. More and more instructors are "docking" students/trainees for absences in the same manner their wages would be reduced for a similar problem in industry.

Since grading methods vary greatly, it is not possible to include detailed evaluation techniques in this manual.

The material in the following sections may be of more concern on the secondary school level than the post-secondary level.

1.7 MAINTAINING DISCIPLINE

Discipline is one of the primary responsibilities of the teacher/instructor. Good class discipline can make teaching a pleasure while poor discipline can make teaching agony.

It would be foolish to assume that there is a set of rules, magic formulas, or chants that will guarantee a teacher/instructor good discipline and respect. There are however, ideas that have been successfully used by many teachers/instructors. Remember, these are only recommendations. It is necessary to work at maintaining good discipline every minute you are in the classroom. There is no one correct way to do it. What works today may not work tomorrow.

- The purpose of discipline is correction. Discipline is *not* chastisement. Discipline is systematic training for improvement of a person's attitude and action.
- Make reprimands with justice and tact. When you are angry or not feeling well, it is wise to refrain from any drastic action until you have the opportunity to review the situation in a better state of mind.
- Be consistent in disciplinary actions.
- Consider the student's mental and physical condition.
- Seek the actual cause of their poor work or attitude.
- Never be influenced by a student's reputation.
- Control order through an interest in your work. Students/trainees sense when you do not like your job.
- Provide sufficient equipment and working materials.
- Keep machines and equipment in good working condition.
- Maintain a clean and attractive shop/lab.

- Be sure all student's can see and hear you when you present a lesson.
- Handle all disciplinary cases yourself whenever possible. A good teacher/instructor is one who seldom has to call on higher authority to maintain discipline in his/her class. Report only major infractions to higher authority. Be able to document problems with complete and up-to-date records.
- Stop problems at their origin.
- Make only necessary rules and *enforce them every day.*
- Avoid assigning schoolwork as punishment. The entire class should not be disciplined for the acts of an individual.
- Make disciplinary actions fit the deeds.
- *Never* use abusive language and/or profanity.
- Grade papers and tests so they can be returned the next class session.
- Have work prepared for the substitute when you must miss a class. Student/trainee time should not be wasted and you will know where to start when you return to class.

Discipline requires more than the few suggestions listed above. Talk to your principal/director on what their ideas are for discipline and what support you can expect from them should a problem arise that you are not able to handle.

1.8 PREPARING FOR THE SUBSTITUTE TEACHER

Substitute teachers have one of the most difficult jobs in education. Some of us expect a substitute teacher to do things we cannot do ourselves. For example, they are to keep a class under control with a lesson plan as brief as, "Read Chapters 1, 2, and 3 and answer the questions at the end of the chapter."

The instructor who maintains a well organized machining technology shop/lab will provide carefully planned material for the substitute teacher. Advanced preparation for instructor absence will permit the program to continue with minimum interruption.

Materials and information that will be required by the substitute teacher include:

- A seating chart and attendance book for each class.
- The location of keys.
- Brief descriptions of the daily routine and the student/trainee organization.

- Procedure for distributing equipment and materials.
- The procedure for cleanup and dismissal.

The operation of equipment should not be allowed.

The best plans are of little value if the substitute cannot locate them. To avoid such a situation, assign a specific place for storing the plans. It is recommended that several instructors in the shop/lab area be informed where the plans are stored.

Request that the substitute teacher leave a report on accomplishments and/or problems for each class. The substitute's report should include the material covered, work completed, and class conduct. The substitute should include the strengths and weaknesses of materials prepared on his/her behalf as well as any suggestions for improvement.

1.9 VARYING STUDENT/TRAINEE ABILITIES

Students/trainees entering your classroom will undoubtedly possess a variety of skills. Some may possess basic metalworking skills while others may not. Students will also have a wide variety of career goals. Some students/trainees plan to attend college and become engineers in the field of metals technology while others will attend community college programs to become technicians. Still others may want to enter the trade or enlist in the Armed Forces as specialists in machining technology.

As with any field of study, the level of student/trainee comprehension and performance will vary within each class. Achievement in machining technology depends not only on a person's communication skills and dexterity, but their ability to visualize solutions to problems. You must recognize the degree to which your students/trainees possess these skills. Different teaching methods are required for students/trainees with different ability levels. Following are some suggestions that may be helpful in serving all students/trainees, especially those with special needs.

1.9.1 Identifying Students/Trainees with Special Needs

Ideally, special needs students/trainees will be identified by the school psychologist. The school nurse should also provide information regarding students/trainees who, because of the medication they are taking, or because of other physical disabilities, should not be permitted to operate machinery. *Actions of some of these students/trainees may endanger themselves or other people in the class. Work out a plan with your principal, supervisor, or director in advance on how this type of a situation should be handled.*

Characteristics of special needs students/trainees are included in the following list. An infrequent occurrence does not denote that someone is a special needs student/trainee.

- Impaired speech, hearing, and/or vision.
- Inability to communicate effectively (reading, writing, oral expression, and questioning).
- Short attention span or lack of interest.
- Lack of motivation.
- Lack of self-confidence.
- Poor computational skills.
- High rate of absenteeism.
- Frequent disruptive behavior.

Mainstreaming

In the past, most special needs students were segregated into separate rooms for instruction. The rooms were generally staffed with instructors trained to work with special needs students. The instructors, however, were not well-trained in vocational-technical disciplines. Mainstreaming is a popular method of integrating special needs students into vocation-technical programs.

A variety of special needs students are commonly found in any classroom. Some may have limited mobility while others may have vision or hearing problems. Always remember that a special needs student needs to be challenged just like any other student. The type of challenge may be based on the student/trainee's potential. Of course, this may require modification in teaching methods.

When working with hearing-impaired students/trainees, talk in a normal voice. Face the class, speak distinctly, and pronounce words clearly. With this approach, a student/trainee with lip reading ability will find it easier to understand the lesson. At the same time, you will be heard more clearly by the rest of the class.

A vision-impaired student/trainee should be allowed to explore the shop/lab with a student companion. The exploration will allow the student to become familiar with locations of equipment and other items. Seat the student toward the front of the class. It is not necessary,

however, to seat the student in the first row. When lecturing, be specific in your references. Pay close attention to the needs of the student/trainee.

There is also the possibility that there will be students/trainees for whom English is not a first language. Speak slowly and clearly and encourage them to communicate as best they can in English. Help them develop their speaking and writing skills so they will feel more comfortable and achieve a higher level of comprehension.

1.9.2 Identifying the Gifted Student/Trainee

The gifted student/trainee may be more difficult to recognize than the special needs student/trainee. In some cases, an instructor is informed of a gifted student by a counselor. In many cases, however, gifted students/trainees have not been identified. Vocational/technical instructors should be familiar with gifted students/trainee's characteristics and modify teaching methods accordingly. Common characteristics include:

- More inclined to question or comment on the subject.
- Consistently scores high on exams and performance activities.
- Consistently completes work ahead of others and requests additional assignments.
- May seem bored, restless or disinterested.
- Displays more interest or a longer attention span.
- Consistently pays close attention to detail on procedures or techniques that are generally taken for granted.

Gifted students can be motivated and challenged by being assigned work with greater complexity, asked to assist slower students/trainees, pressed into service as assistants in the classroom and shop/lab, given problem-solving activities that involve creative thinking, and given special topics to research and report to the class.

1.10 ADDITIONAL RESOURCES

A variety of supplemental materials may be used to allow the students/trainees to further develop their interest in machining technology. Not all learning occurs in the classroom. Students/trainees should be encouraged to read supplemental material, participate in field trips, and join organizations in the field. When possible, outside activities should be followed up by some means of discussion and/or evaluation.

1.10.1 Field Trips

Field trips can be an excellent way to motivate students. Every community, large or small, offers opportunities in the metalworking fields. Field trips should be carefully planned and arrangements made ahead of the visit.

Students/trainees should receive a briefing prior to the trip. They will then know what to look for and can better understand what they observe. If possible, have a representative from the shop/company visit your classroom and give a presentation prior to the field trip.

A follow-up evaluation and discussion is a must after a field trip. Often having students write a short theme on what was observed can be helpful. Such a report will also improve a student/trainee's writing ability.

1.10.2 Vocational-Technical Fairs

Many schools, vocational-technical centers, and community colleges use fairs to motivate students. Students should also be involved when planning such an exhibition. The display may be in the shops/labs, local mall, or some other location.

Machining technology students/trainees may want to display completed assignments and give demonstrations.

1.10.3 Vocational-Technical Organizations

Encourage students/trainees to get involved in recognized organizations such as the Vocational Industrial Clubs of America (VICA). These organizations promote vocational-technical education and provide worthwhile experiences for the students/trainees.

These organizations also strive to develop character, as well as leadership abilities. Strong emphasis is placed on the respect for the dignity of work, high standards in ethics, workmanship, scholarship, and safety.

1.10.4 Awards Dinner

An annual awards dinner is an excellent way to reward student/trainee achievement. It also offers an opportunity to gain recognition of your program.

Invite parents, administrators, prominent citizens, future employers, and the press to the event.

1.11 SHOP/LAB MANAGEMENT

Good shop/lab management starts several days *before* the first class session. It should include conservation (time, effort, and energy), plus student/trainee safety, so the major endeavor can be directed toward improving instruction, which in turn, enhances learning.

Shop/lab management also requires some paperwork. The better you prepare for the first class session, the more time you will have to take care of the many minor problems that crop up when the students/trainees arrive. Get all of the forms you plan to use made up and reproduced in advance. Be sure they are readable, and prepare enough for all of your classes.

Accident Reports

In many learning situations, it is mandatory that each accident or injury be reported. Fill out a report immediately in those situations. Have the injured person report as soon as possible to the infirmary. Check with your principal, supervisor, or director for specific instructions on filling out the required forms.

1.12 TEACHER/INSTRUCTOR RESPONSIBILITY

Most states have laws relating to safety requirements in various areas of educational programs. Contact the necessary authorities to check your state's laws and local requirements.

In machining technology they usually relate to mandatory eye protection for special areas of the shop/lab. Such as when:
- Turning, shaping, cutting, or stamping solid materials with hand tools and/or power driven tools, machines, or other equipment.
- Casting, welding, and heat treating.
- Handling caustic materials.

While not usually mandated by state law, some systems have established policies on keeping certain problems under control or eliminating them as a possible source of student/trainee injury. They include:
- Appropriate footwear must be worn in metalworking areas.
- Long hair must be contained while operating power equipment.
- Use of drugs (medication and/or illegal) that impair muscular coordinating, spatial perception, etc.

You will find it to your advantage to know and understand the laws and requirements affecting your program.

Safety of students/trainees is paramount. Dangers from toxic gases, lack of oxygen, flying particles, sharp metal edges, hot metal, electrical shock, and falling objects are ever present. A strong, firm safety policy is mandatory for both students and technicians. Safety warning and cautions (*printed in red*), are located throughout the *Machining Fundamentals* textbook. These warnings are *not* exhaustive.

The office of Occupational Safety and Health Administration (OSHA) requires assurance that working conditions are kept healthy as far as vapors, fumes, temperature, dust, air contaminants, and noises are concerned.

Safety should be a top priority for everyone. It is the best and most efficient way to do any task. Common sense safety rules include the following:

- Keep work stations clean and orderly.
- Wear correct clothing for the job. Students should wear gloves when handling substances of a corrosive or sharp nature. Protect clothing by wearing an apron. Loose clothing can get caught, and a careless movement can mangle fingers. Advise students/trainees that they must wear safety shoes, hard hats, and hearing protection.
- Respect fire.
- Use proper lighting.
- Keep aisles and exits clear.
- Avoid electrical shocks. Be sure the electrical circuit is locked open before handling electrical equipment. Avoid standing or sitting in damp or wet places when doing electrical work. Never allow students to work with wet hands. The instructor should check all student-installed electrical service before allowing it to be "plugged-in" to the electrical circuit.
- Report all accidents and injuries.
- Wear safety glasses, goggles, and face shields when appropriate. Proper eye protection should be worn at all times in the shop. Goggles worn in the shop/lab should meet USASI Safety Code Z17.1-1968 (United States of America Standards Institute). When not being used, goggles should be stored in a cabinet and exposed to a germicidal lamp to destroy bacteria.
- Use adequate respiratory protection. Proper ventilation is vitally important in any shop or lab environment. Toxic vapors and fumes must be minimized. Exhaust blowers, along with makeup air units, should be used.
- Maintain and store equipment, tools, and supplies correctly. Never operate machinery without *all* safety guards in place.
- Use proper tools and procedures.
- Disconnect all power when tools or equipment are not in use.

The Occupational Safety and Health Act (OSHA) establishes standards affecting the safety of workers in any occupation.

OSHA is a federal law that mandates most safety recommendations published by organizations. Graduates who find employment in industry should know and practice these regulations. The act is being enforced by the U.S. Department of Labor.

A copy of the act should be available to all instructors. Copies of the Federal Act are available from the Superintendent of Documents, U.S. Government Printing Office, Washington, DC, 20402. Copies of the state act are also available from state Departments of Safety.

Job Safety and Health, published by the U.S. Department of Labor, Occupational Safety and Health Administration, may be used in a classroom or laboratory for discussing safe working conditions. It is available from the U.S. Government Printing Office.

The National Safety Council, 1121 Spring Lake Drive, Itasca, Illinois, 60143-3201, publishes safety education data sheets and posters that cover safety in the home, on the job, and in various other locations. A list of the available safety education data sheets can be obtained by writing the council at the above address.

Most cities have codes concerning the installation and operation of refrigeration and air conditioning equipment. Students should be acquainted with the code or codes for their city.

1.12.1 Other Areas of Responsibility

We live in a litigious society. Because of the nature of working in a machine shop/lab, industrial-technical instructors must be acquainted with situations that might cause legal actions to be brought against them and the institution where they teach.

Often times, an act of negligence is brought about by ignoring basic safety rules or not using common sense. The following list has been provided to enlighten your sense of responsibility. It is not all-inclusive. If you have serious concerns in regards to legal matters and liabilities, contact the appropriate department(s) of your school system.

- *Perform all machining operations in a proper and safe manner.* Failure to employ the proper use of safety guards, safety goggles or glasses, protective clothing, and proper procedure can endanger both you and your students/trainees. If students/trainees are taught improper and unsafe practices, they will use improper and unsafe practices.
- *Keep all circumstances under your control.* Although you may perform a demonstration with the utmost care and precaution, if the area is disorganized or dirty (i.e., oil or cutting fluid on the floor), it is likely that someone may be injured.
- *Do not entrust tools or machine operation to a student/trainee that you know is likely to inflict intended harm upon others.* If someone has been unruly or has threatened to harm fellow classmates, report the incident to the appropriate authorities. If you believe someone is a threat to his/her classmates, take the appropriate steps for their removal.
- *Safety cannot be overemphasized.* Remind students on a daily basis to stay alert and pay attention to danger zones.
- *Give students/trainees adequate preparation before allowing them to perform any machining operation.* Repeat procedures and safety practices in lectures, during demonstrations, and while a student performs the operation.
- *Always keep tools and machines in good working condition.* Do *not* allow students to work on machines that are in disrepair. If you are unable to repair the problem yourself, change the lesson plans or assignment until the repairs can be made. Inspect tools and machines on a daily basis and encourage your students/trainees to do the same.

As stated earlier, this is not an all-inclusive list. Adequate preparation, emphasis on safety, proper maintenance, and the application of common sense will help maintain a safe and productive shop/lab.

1.13 KEEPING UP-TO-DATE

As with all technical areas, there are constant changes and improvements in machining technology. The following material may help keep your program in tune with advances and changes in the metalworking field.

1.13.1 Advisory Council

The establishment of an advisory council is usually required when program financial aid is received from the state and/or federal government. The council should be composed of people who are actively engaged in machining technology.

The function of an advisory council is to advise instructors on the needs and trends in the industry. They can also help keep programs current, and provide contacts for placement of students/trainees completing the program.

1.13.2 Additional Means for Keeping Up-to-Date

Since there is constant change and improvements in all technical areas of machining technology, there are, in addition to the advisory council, other ways to keep up-to-date with these changes.

To keep pace with change, instructors should subscribe to and read monthly trade magazines and journals; visit modern machine shops and talk with management and personnel; attend seminars sponsored by the various machining technology vendors and trade associations; and attend trade shows and conventions.

Some employers offer special sabbatical leaves to allow instructors to work in the field for a specified period of time, so as to gain experience in the latest techniques available.

Summer employment in a modern shop would also be helpful. Summer session at the college level can be beneficial in improving teaching techniques and in developing a better understanding of the overall educational picture.

Keeping abreast with modern developments in machining technology is mandatory if you want to offer a course that will equip your students with the skills, knowledge, and attitudes needed today.

1.14 RESOURCE MATERIALS

There is an immeasurable amount of resource material available on the metalworking industry. Maintain a library if possible or request that the library carry a selection of textbooks and publications on the field. The following books, catalogs, manuals, and periodicals are only a brief listing of available material.

1.14.1 Reference Books

Amman, Jost, *The Book of Trades* (Historical): Dover Publications, (Original publication date 1658).

Barsamian, Michael, and Gizelbach, Richard A., *Machine Trades Print Reading:* Goodheart-Willcox Publisher.

Biekert, Russell, *CIM Technology:* Goodheart-Willcox Publisher.

Brandt, Daniel A., *Metallurgy Fundamentals:* Goodheart-Willcox Publisher.

Brown, Walter C., *Basic Mathematics:* Goodheart-Willcox Publisher.

Brown, Walter C., *Print Reading for Industry:* Goodheart-Willcox Publisher.

Diderot, Denis, *A Dideroit Pictorial Encyclopedia of Trades and Industry,* Vol. 1 and 2 (Historical): Dover Publications, (Original publication date 1752).

Duenk, Lester G., Editor, *Improving Vocational Curriculum:* Goodheart-Willcox Publisher.

DuVall, J. Barry, *Contemporary Manufacturing Processes:* Goodheart-Willcox Publisher.

Gradwell, John; Welch Malcom; and Martin, Eugene, *Technology Shaping Our World:* Goodheart-Willcox Publisher.

Green, Robert E., Editor, *Machinery's Handbook:* Industrial Press Inc.

Littrell, J. J., *From School to Work:* Goodheart-Willcox Publisher.

Madsen, David A., *Geometric Dimensioning and Tolerancing:* Goodheart-Willcox Publisher.

Masterson, James; Towers, Robert; and Fardo, Stephen, *Robotics Technology:* Goodheart-Willcox Publisher.

Phagan, R. Jesse, *Applied Mathematics:* Goodheart-Willcox Publisher.

Walker, John R., *Modern Metalworking:* Goodheart-Willcox Publisher.

Wanat, John H.; Pfeiffer, E. Weston; and Van Gulik, Richard, *Learning for Earning:* Goodheart-Willcox Publisher.

Wilson, Bruce A., *Design Dimensioning and Tolerancing:* Goodheart-Willcox Publisher.

1.14.2 Catalogs and Manuals

American National Standards Institute (ANSI)
1430 Broadway
New York, NY 10013
(Catalog of standards and price list)

The Association of Manufacturing Technology
7901 Westpark Drive
McLean, VA 22102-4206
(List of machine tool manufacturers)

Dover Publications, Inc.
31 East 2nd Street
Mineola, NY 11501
(Catalog on metalworking history, etc.)

Hanser Gardner Publications
6915 Valley Avenue
Cincinnati, OH 45244-3029
(Catalog of technical books and audio visual materials)

The Industrial Press
93 Worth Street
New York, NY 10013
(Machinery's Handbook)

The M.I.T Press
Massachusetts Institute of Technology
Cambridge, MA 02142

National Tool & Machining Association
9300 Livingston Road
Fort Washington, MD 20744
(Catalog of technical books and audio visual materials)

U.S. Bureau of Labor Statistics
Government Printing Office
200 Constitutional Ave., NW
Washington, DC 20210
(Occupational Handbook)

Catalogs of precision tools, machine tools, etc. Journals of various industrial and trade associations.

1.14.3 Periodicals

American Machinist
Penton Publishing
1100 Superior Avenue
Cleveland, OH 44114

Automation News
155 E. 23rd Street
New York, NY 10010

CAD/CAM & Robotics
Kerrwil Publications Ltd.
501 Oakdale Road
Downsview, ON, Canada M3N 1W7

Cutting Tool Engineering Magazine
400 Skokie Blvd., Suite 395
Northbrook, IL 60062

The Home Shop Machinist
The Village Press
2779 Aero Park Drive
Traverst City, MI 49686

Industrial Education
Cummins Publishing Company
26011 Evergreen Road
Southfield, MI 48076

Industrial Machinery Digest
One Chase Corporate Drive #300
Hoover, AL 35244

Machine Design
1100 Superior Avenue
Cleveland, OH 44144

Metalfax
29100 Aurora Road
Solon, OH 44139

Metalworking Digest
1350 East Touhy Avenue
Des Plaines, IL 60017

Modern Machine Shop
6600 Clough Pike
Cincinnati, OH 45244-4090

Tooling and Production
29100 Aurora Road, Suite 200
Solon, OH 44139

1.14.4 Agencies and Associations

American National Standards Institute (ANSI)
1430 Broadway
New York, NY 10018

American Vocational Association
1410 King Street
Alexandria, VA 22314

International Technology Education
 Association (ITEA)
1914 Association Drive
Reston, VA 22091

National Association of Industrial
 Technology (NAIT)
3300 Washenaw Avenue, Suite 220
Ann Arbor, MI 48104-4200

National Institute for Metalworking Skills
2209 Hunter Mill Road
Vienna, VA 22181

Vocational Industrial Clubs of America (VICA)
Box 30
Leesburg, VA 22075

1.14.5 Audiovisual Materials

A variety of audiovisual materials is available for use in machining technology. Contact the following companies and associations for listings of available materials.

DCA Educational Products, Inc.
1814 Kellers Church Road
Bedminster, PA 18910

Hanser Gardner Publications
6915 Valley Avenue
Cincinnati, OH 45244-3029

Minnesota Mining and Mfg. Co., 3M Center
Visual Systems Division
Austin, Texas, 78769

National Tooling & Machining Association
9300 Livingston Road
Fort Washington, MD 20744

L.S. Starrett Company
121 Crescent Drive
Athol, MA 01331

Sterling Educational Films
241 E. 34th Street
New York, NY 10016

1.14.6 On-Line Resources

Many resources for Machining Technology and education are available over the information superhighway. They are sponsored by corporations, private organizations, and individuals.

On-line addresses for information on all areas of machining technology can be found in trade magazines and journals. Using the Internet can keep you current on new machine tool developments, and machining and manufacturing techniques.

Most of the sites provide information free of charge or for a minimal fee. Since the information superhighway is constantly expanding, the addresses for some of the following sites may have changed since the publication of this manual. The following is only a sampling of companies with on-line sites:

Cincinnati Milacron (CNC machine tools)
www.milacron.com

Mastercam (software)
www.mastercam.com

Modern Machine Shop Magazine
www.gardnerweb.com/mms

Nikon
(optical comparators and CNC
Video measuring systems)
www.nikonusa.com

Sharnoa Corp. (CNC machine tools)
www.sharnoa.com

South Bend Lathe Corp. (machine tools)
www.southbendlathecorp.com

GOODHEART-WILLCOX WELCOMES YOUR INPUT

If you have comments, corrections, or suggestions regarding the textbook or its supplements, please send them to:

Managing Editor
Goodheart-Willcox Publisher
18604 West Creek Drive
Tinley Park, IL 60477-6243

NOTES

Meeting Certification Standards

As certification standards are fast becoming employment requirements, it is vital that vocational/technical programs cover the skills required by industry. To help plan your program using the *Machining Fundamentals* text, the following information, correlation chart, and drawings have been included.

The National Tooling and Machining Association (NTMA) has developed a three level program *(Duties and Standards for Machining Skills)* of National Skill Standards for the Machining Industry. The goal of the NTMA was to establish world-class standards reflecting industry skill requirements. The standards are performance based and at present are voluntary.

The standards will provide a method through certification and training for individual workers to receive recognition and reward for their abilities. The standards will help employers identify training needs and evaluate job applicants fairly.

Elements of all three standards often appear in training programs depending on capabilities available and in certain programs where the concentration of work is in machine specific operations. Due to the scope of *Machining Fundamentals*, only Level I will be covered in this manual. For additional information regarding the standards and testing programs, contact the National Tooling and Machining Association (NTMA), 9300 Livingston Road, Fort Washington, MD 20744.

Level I Machining Responsibilities

According to the NTMA, an individual with Level I Machining Skills is a skilled machine operator or technician who has demonstrated competence in three major areas of responsibility: basic bench operations, basic metal cutting operations, and basic inspection and quality assurance functions.

This individual can perform these responsibilities in both single and multiple part production. No supervision or training responsibilities of other operators or other production workers is assumed at Level I.

Most Level I skills can be met in six months to one year of education and training, depending on prior manufacturing experience, basic academic skills, mechanical aptitude, and the availability of laboratory-based training. This training could be given in a high school or community college vocational/technical education program, apprenticeship program, formal company training program, or structured on-the-job training.

Each skill set is based on the most important responsibilities that workers are expected to perform and is modular in design—student/trainees, workers, or employers select those from the different skill sets that best meet their career direction or job requirement.

Level I machining responsibilities typically include the ability to: (Note: These are not the standards.)

Bench operations:
- Select and use hand tools.
- Perform basic, routine layout.
- Read and comprehend information on orthographic prints and job process sheets for manufacturing operations.
- Deburr.
- Perform hand fitting and minor assembly.
- Perform bench cutting tasks such as sawing, reaming, and tapping.

- Perform basic, routine preventive maintenance.
- Perform basic housekeeping responsibilities.

Metal cutting operations:
- Identify basic metallic and non-metallic materials.
- Identify and use most accessories and tooling for machining operations.
- Choose an appropriate speed and feed for a given operation.
- Perform basic process planning, setup, and operation of common classes of machine tools such as turning, milling, drilling, or surface grinding machines.
- Select and use cool coolants appropriately.
- Make suggestions for improving basic machining operations within a structured improvement process.
- Be competent in all safety procedures for all machining operations and material handling and disposal within their responsibility.

Inspection and quality assurance responsibilities:
- Use basic precision measurement tools.
- Follow an inspection process plan.
- Perform basic quality assurance responsibilities for both single and multiple part production including statistical process control.

Other competency areas:
- Follow standardized work procedures in a limited range of standardized work contexts under direct supervision.
- Be competent in all basic aspects of seeking and maintaining employment in the metalworking industry.

Duty Framework for Machining Skills—Level I

The following is a complete list of the duty areas and titles that comprise the NTMA's Level I machining skills. This outline can be used to help ensure that your program includes all of the areas identified.

1. **Job Planning and Management**
   1.1   Job Process Planning
2. **Job Execution**
   2.1   Manual Operations Benchwork
   2.2   Manual Operations Layout
   2.3   Turning Operations, Between Centers Turning

   2.4   Turning Operations, Chucking
   2.5   Power Feed Milling
   2.6   Vertical Milling
   2.7a Grinding Wheel Safety
   2.7b Surface Grinding
   2.8   Drill Press Operations
   2.9   Power Saws
3. **Quality Control and Inspection**
   3.1   Part Inspection
   3.2   Process Control
4. **Process Adjustment and Control**
   4.1   Process Adjustment, Single Part Production*
   4.2   Participation in Process Improvement*
5. **General Maintenance**
   5.1   General Housekeeping and Maintenance
   5.2   Preventive Maintenance*
   5.3   Tooling Maintenance*
6. **Industrial Safety and Environmental Protection**
   6.1   Machine Operations and Material Handling*
   6.2   Hazardous Materials Handling and Disposal*
7. **Career Management and Employment Relations**
   7.1   Career Planning
   7.2   Job Applications and Interviewing*
   7.3   Teamwork and Interpersonal Relations*
   7.4   Organizational Structures and Work Relations*
   7.5   Employment Relations*

*   *These duty titles are beyond the scope of the text and have been omitted from the correlation chart.*

In addition to basic machining skills, students/trainees must also master basic communication, math, measurement, and drawing skills. The basic skills listed below are correlated to the duty areas by number.

1. **Written and Oral Communications**
   1.1   Reading
   1.2   Writing
   1.3   Speaking
   1.4   Listening
2. **Mathematics**
   2.1   Arithmetic
   2.2   Applied Geometry
   2.3   Applied Algebra
   2.4   Applied Trigonometry
   2.5   Applied Statistics

3. **Decision Making and Problem Solving**
   3.1   Applying Decision Rules
   3.2   Basic Problem Solving

4. **Group Skills and Personal Qualities**
   4.1   Group Participation and Teamwork
   4.2   Personal Qualities

5. **Engineering Drawings and Sketches**
   5.1   Standard Orthographic Prints
   5.2   GDT Orthographic Prints

6. **Measurement**
   6.1   Basic Measuring Instruments
   6.2   Precision Measuring Instruments
   6.3   Surface Plate Instruments

7. **Metalworking Theory**
   7.1   Cutting Theory
   7.2   Tooling
   7.3   Material Properties
   7.4   Machine Tools
   7.5   Cutting Fluids and Coolants

Prints for Developing Level I Skills

The prints included on pages 27–37 of this resource were developed by the author in the attempt to fulfill requirements for the Level I assignments. The prints are correlated to the duty titles by the drawing numbers. Instructors are encouraged to prepare job assignments of their own, or use industrial drawings that will offer the same challenges.

The tolerances noted on the drawings are those recommended by The Metalworking Industry Skills Standards Board.

Although 1018 H.R. steel has been selected for the assignments, cold finished steel, steels of other composition, or other metals can be used.

The following list of materials and tools (identified by duty title number) has been included for your convenience in organizing the materials required for each assignment.

Angle plate, 2.2, 2.7b, 2.8, 3.1
Appropriate population of product matching print specifications (broken up into packages matching requirements of sampling plan), 3.2
Arbor press, 2.1
Ball-peen hammer, 2.2, 2.5, 2.6, 2.8
Bench vise (4"), 2.1
Blade(s), (power saw), 2.9
Boring bar, 2.4
Boring tool (capable of boring to a square shoulder), 2.4
Brooms, 5.1
Brushes, 5.1

C-clamps, 2.2
Center drills, 2.4, 2.8
Center gage, 2.3, 2.4
Chamfer tools (45°), 2.3, 2.4
Clamps, 2.5, 2.6, 2.7b, 2.8, 3.1
Combination drill and countersink, 2.3, 2.6
Combination set, 2.2, 2.3, 2.4, 2.5, 2.6, 2.7b, 2.8, 5.3
Common workbench, 1.1, 2.1, 2.2, 2.3, 2.4, 2.5, 2.6, 2.7a, 2.7b, 2.8, 3.1, 3.2, 4.1, 5.1, 5.3
Composition hammer, 2.7b, 2.8
Coolants, 2.6
Counterbores, 2.8
Countersink, 2.6, 2.8
Cutter (2" or larger diameter) that may be a face mill, 2.5
Cutter adaptors, 2.5, 2.6
Cutters (assorted) , 2.5 , 2.6
Cutting fluids, 2.8, 4.1
Cutting oil, 2.6
Dead center fitted to the spindle taper, 2.3
Depth micrometer set, 2.6, 2.7b
Dial indicator, 2.3, 2.4, 2.5, 2.6, 2.7b, 5.3
Diamond dresser, 2.7a, 2.7b
Dividers (6"), 2.2, 2.8
Drill chucks, 2.3, 2.4, 2.6, 2.8
Drill press (Morse taper #3 spindle capacity or greater; tapping capability), 2.8
Drill vise (6" or greater), 2.8
Drills, 2.4, 2.6, 2.8, 5.3
Edge finder, 2.5, 2.6
Engine lathe (min. 14" 3 30" capacity), 2.3, 2.4
EPA guidelines for hazardous material handling and storage, 6.2
External threading tool matched to profile of thread on turning print, 2.3, 2.4
External undercut tools, 2.3, 2.4
Files, 2.3, 2.4, 2.5, 2.6, 2.7b, 2.9
Fishbone charts, 4.2
Flip charts, 4.2
Forty taper spindle or greater, 2.5
4-jaw independent chuck, 2.3, 2.4
Gage blocks (inspection grade), 3.1
Go/no-go gage for threads, 2.8
Grease, 5.2
Grinding wheels (assorted), 2.7a, 2.7b
Hand tools, 2.7a, 5.2
Hazardous materials, 6.2
   identification instruments, 6.2
   instruments for measuring concentration, 6.2
Inside calipers, 2.4
Inspection tools (basic, fixed, precision, surface plate), 3.1, 3.2, 4.1
Internal threading tool matched to profile of thread on turning print, 2.4

Correlation Chart

The following chart correlates the NTMA's Level I *Duties and Standards for Machining Skills* to the *Machining Fundamentals* text. The duty areas and titles are correlated by chapter and section numbers.

MACHINING SKILLS—LEVEL I ASSIGNMENTS

Duty Area	Duty	*Machining Fundamentals*	Drawing Nos.
1. Job Planning and Management	**1.1 Job Process Planning** Develop a process plan for a part requiring milling, drilling, turning, or grinding. Fill out an operation sheet detailing the process plan and required speeds and feeds.	Chapter 3, 29–54.	B234789
2. Job Execution	**2.1 Manual Operations: Benchwork** Tap holes. Use files, scrapers, and coated abrasives to deburr parts. Use arbor presses to perform press fits. Use bench vises and hand tools appropriately.	Chapter 2, 23–28; Chapter 3, 29–54; Chapter 6, Sections 6.1.1 Vises, 6.7 Hacksaw, 6.8 Files, 6.10 Hand Threading, 6.11 Polishing.	B234790
2. Job Execution	**2.2 Manual Operations: Layout** Lay out the location of hole centers and surfaces within an accuracy of $\pm -.015$.	Chapter 2, 23–28; Chapter 3, 29–54; Chapter 4, 55–80; Chapter 5, 81–90.	B234791
2. Job Execution	**2.3 Turning Operations: Turning Between Centers** Set up and carry out between centers turning operations for straight turning.	Chapter 2, 23–28; Chapter 3, 29–54; Chapter 4, Sections 4.1 The Rule, 4.2 Micrometers, 4.3 Vernier Calipers, 4.4.4 Thread Gages, 4.5 Dial Indicators, 4.6.7 Screw Pitch Gages; Chapter 9, 149–152; Chapter 11, 183–190; Chapter 13; 201–240; Chapter 14, 241–260.	B234792
2. Job Execution	**2.4 Turning Operations: Chucking** Set up and carry out chucking operations for turning.	Chapter 2, 23–28; Chapter 3, 29–54; Chapter 4, 55–80; Chapter 9, 149–152; Chapter 13, 201–240; Chapter 14, 241–260.	B234793
2. Job Execution	**2.5 Power Feed Milling** Set up and operate a horizontal or vertical milling machine using power feeds. Perform routine milling.	Chapter 2, 23–28; Chapter 3, 29–54; Chapter 4, 55–80; Chapter 9, 149–152; Chapter 17, 285–316; Chapter 18, 317–352.	B234794

Duty Area	Duty	*Machining Fundamentals*	Drawing Nos.
2. Job Execution	**2.6 Vertical Milling** Set up and operate vertical milling machines. Perform routine milling and location of hole centers within ±.005″.	Chapter 2, pages 23–28; Chapter 3, 29–54; Chapter 4, 55–80; Chapter 9, 149–152; Chapter 17, 285–317.	B234795
2. Job Execution	**2.7a Surface Grinding: Grinding Wheel Safety** Ring test grinding wheels, perform visual safety inspection, mount and dress a grinding wheel in preparation for surface grinding.	Chapter 2, 23–28; Chapter 19, 149–152.	
2. Job Execution	**2.7b Surface Grinding: Horizontal Spindle Reciprocating Table** Set up and operate manual surface grinders with a 10″ and smaller diameter wheel. Perform routine surface grinding, location of surfaces, and squaring of surfaces. Perform wheel dressing.	Chapter 2, 23–28; Chapter 3, 29–54; Chapter 4, 55–80; Chapter 9, 149–152; Chapter 19, 353–382.	B234796
2. Job Execution	**2.8 Drill Press** Set up and operate drill presses. Perform routine drill press operations.	Chapter 2, 23–28; Chapter 3, 29–54; Chapter 4, 55–80; Chapter 9, 149–152; Chapter 10, 153–182.	B234797, B234798
2. Job Execution	**2.9 Power Saw Operations** Set up and operate power saws for cutoff operations.	Chapter 2, 23–28; Chapter 3, 29–54; Chapter 4, Section: 4.1 The Rule; Chapter 12, 191–200.	B234799
3. Quality Control and Inspection	**3.1 Part Inspection** Develop an inspection plan and inspect simple parts using precision tools and techniques. Prepare reports on the compliance of parts.	Chapter 3, 29–54; Chapter 4, 55–80; Chapter 23, 435–450.	
3. Quality Control and Inspection	**3.2 Process Control** Follow a sampling plan. Inspect the samples for the required data. Enter the data on appropriate charts. Graph the data. Respond to the warning conditions indicated by process charts.	Chapter 3, 29–54; Chapter 4, 55–80; Chapter 23, 435–450.	
5. General Maintenance	**5.3 Tooling Maintenance** Keep the duty station clean and safe for work. Keep the tools, workbenches, and manual equipment clean, maintained, and safe for work.	Chapter 2, 23–28; Chapter 4, 55–80; Chapter 10, Section: 10.8 Sharpening Drills; Chapter 11, 183–190; Chapter 13, Section: 13.6 Cutting Tools and Tool Holders; Chapter 17, Section: 17.4 Milling Cutters; Chapter 19, Sections 19.8 Universal Tool and Cutter Grinder and 19.9 Tool and Cutter Grinding Wheels.	
7. Career Management and Employment Relations	**7.1 Career Planning** Develop and explain a short-term career plan and resume.	Chapter 30, 547–556.	

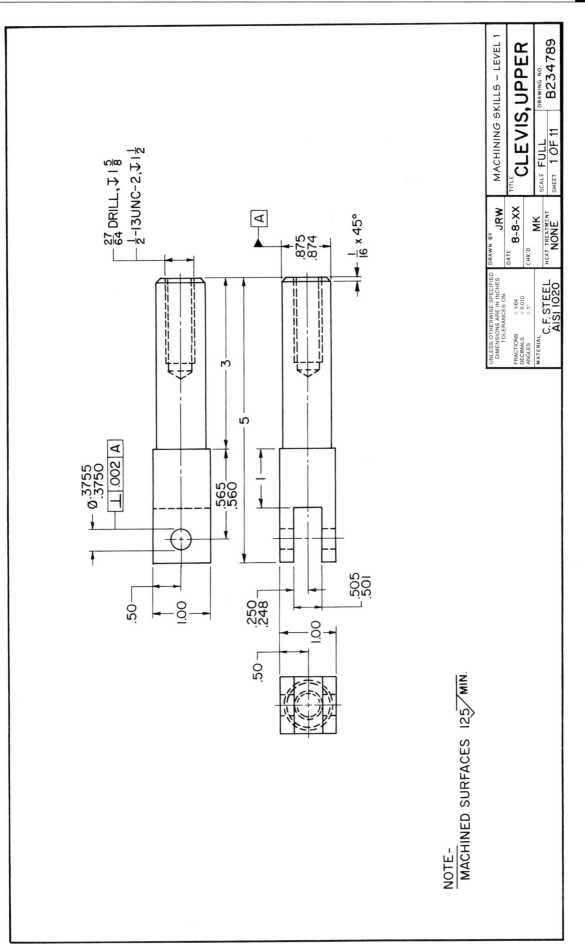

NOTE-
MACHINED SURFACES 125 $\sqrt{}$ MIN.

UNLESS OTHERWISE SPECIFIED DIMENSIONS ARE IN INCHES TOLERANCES ON	DRAWN BY JRW	MACHINING SKILLS – LEVEL 1
	DATE 8-8-XX	TITLE CLEVIS, UPPER
FRACTIONS ± 1/64 DECIMALS ± 0.010 ANGLES ± 1°	CHK'D MK	DRAWING NO. B234789
MATERIAL C.F. STEEL AISI 1020	HEAT TREATMENT NONE	SCALE FULL SHEET 1 OF 11

Drawing callouts:
- $\frac{27}{64}$ DRILL, ⌴ $1\frac{5}{8}$
- $\frac{1}{2}$-13UNC-2, ⌴ $1\frac{1}{2}$
- ∅ $\frac{.3755}{.3750}$
- ⊥ .002 A
- $\frac{.565}{.560}$
- .50
- 1.00
- 3
- 5
- A
- $\frac{.875}{.874}$
- $\frac{1}{16}$ x 45°
- 1
- $\frac{.505}{.501}$
- $\frac{.250}{.248}$
- 1.00
- .50

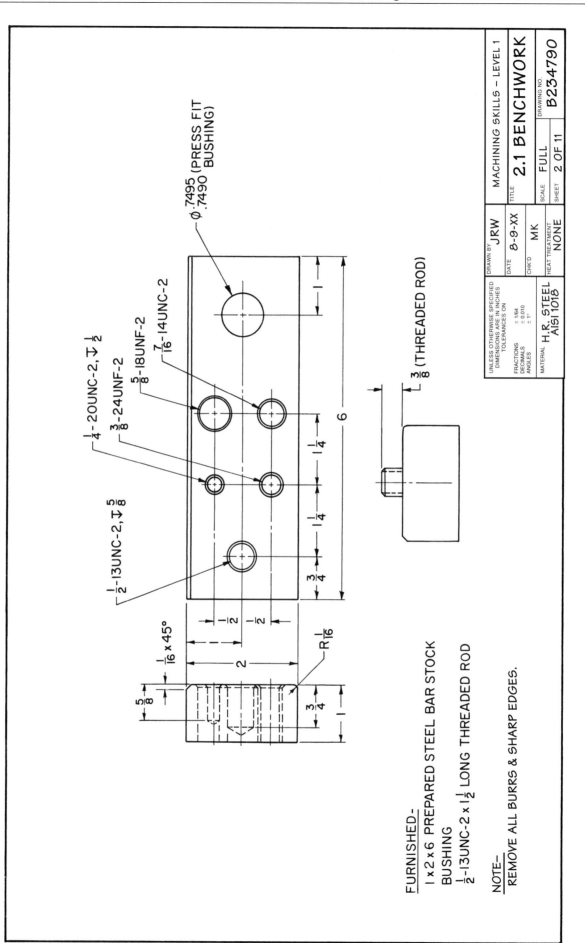

FURNISHED-
1 x 2 x 6 PREPARED STEEL BAR STOCK
BUSHING
$\frac{1}{2}$-13UNC-2 x $1\frac{1}{2}$ LONG THREADED ROD

NOTE-
REMOVE ALL BURRS & SHARP EDGES.

ϕ .7495 (PRESS FIT
.7490 BUSHING)

$\frac{1}{4}$ - 20UNC-2, ⊽ $\frac{1}{2}$

$\frac{3}{8}$-24UNF-2

$\frac{5}{8}$-18UNF-2

$\frac{7}{16}$-14UNC-2

$\frac{1}{2}$-13UNC-2, ⊽ $\frac{5}{8}$

$\frac{3}{8}$ (THREADED ROD)

UNLESS OTHERWISE SPECIFIED DIMENSIONS ARE IN INCHES TOLERANCES ON	DRAWN BY JRW	MACHINING SKILLS – LEVEL 1
FRACTIONS ± 1/64 DECIMALS ± 0.010 ANGLES ± 1°	DATE 8-9-XX	TITLE **2.1 BENCHWORK**
	CHK'D MK	
MATERIAL H.R. STEEL AISI 1018	HEAT TREATMENT NONE	DRAWING NO. **B234790**
		SCALE FULL SHEET 2 OF 11

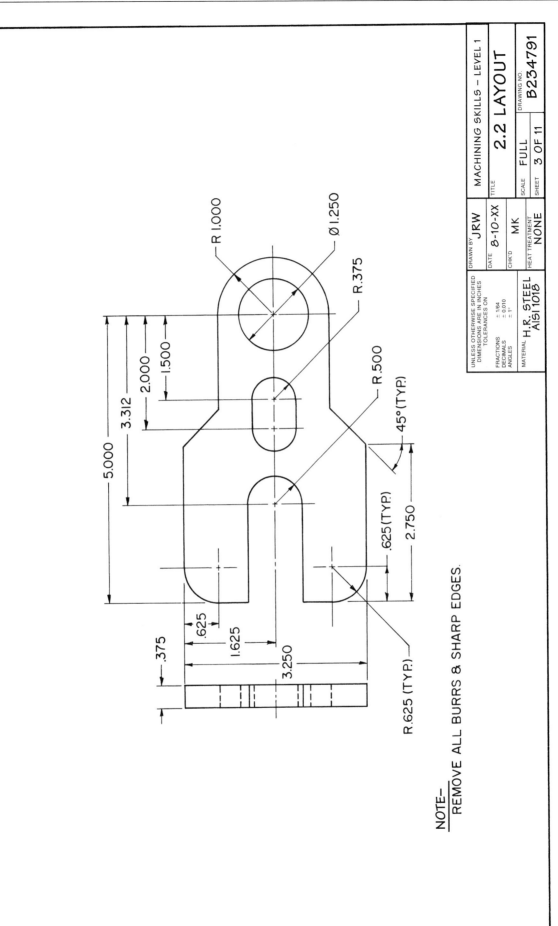

NOTE–
REMOVE ALL BURRS & SHARP EDGES.

R 1.000
Ø 1.250
R.375
R.500
45° (TYP.)
2.750
.625 (TYP.)
R.625 (TYP.)
1.500
2.000
3.312
5.000
.375
.625
1.625
3.250

UNLESS OTHERWISE SPECIFIED DIMENSIONS ARE IN INCHES TOLERANCES ON	DRAWN BY JRW	MACHINING SKILLS – LEVEL 1
	DATE 8-10-XX	TITLE 2.2 LAYOUT
FRACTIONS ± 1/64 DECIMALS ± 0.010 ANGLES ± 1°	CHK'D MK	
MATERIAL H.R. STEEL AISI 1018	HEAT TREATMENT NONE	SCALE FULL DRAWING NO. B234791
		SHEET 3 OF 11

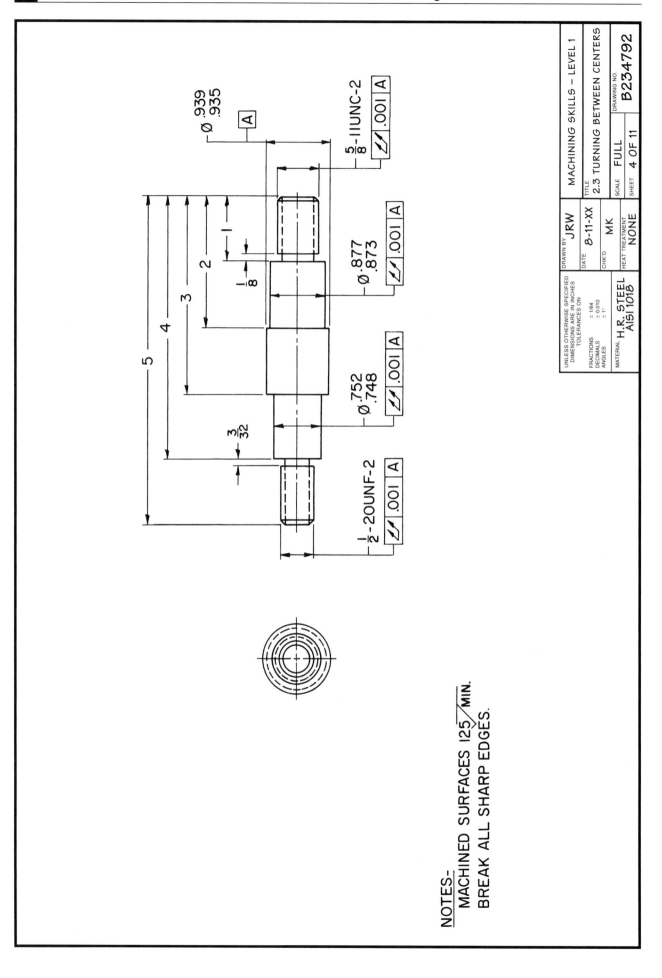

NOTES-
MACHINED SURFACES 125 √ MIN.
BREAK ALL SHARP EDGES.

Ø .939 / .935

⌖ A

5/8-11UNC-2

⟂ .001 A

Ø .877 / .873

⟂ .001 A

Ø .752 / .748

⟂ .001 A

1/2-20UNF-2

⟂ .001 A

UNLESS OTHERWISE SPECIFIED DIMENSIONS ARE IN INCHES TOLERANCES ON		DRAWN BY	JRW	MACHINING SKILLS – LEVEL 1	
FRACTIONS ± 1/64		DATE	8-11-XX	TITLE 2.3 TURNING BETWEEN CENTERS	
DECIMALS ± 0.010		CHK'D	MK		DRAWING NO. B234792
ANGLES ± 1°				SCALE FULL	
MATERIAL H.R. STEEL AISI 1018		HEAT TREATMENT NONE		SHEET 4 OF 11	

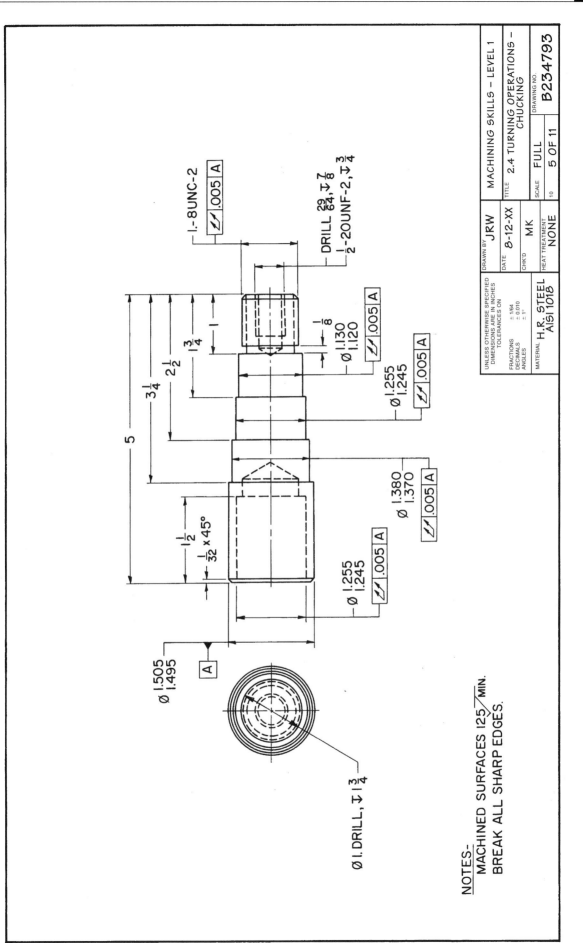

1-8UNC-2 ⊥.005 A

DRILL 29/64, ⊥⅞
½-20UNF-2, ⊥¾

⊥.005 A

Ø1.130/1.120 ⅛

⊥.005 A

Ø1.255/1.245 ⊥.005 A

Ø1.380/1.370 ⊥.005 A

Ø1.255/1.245 ⊥.005 A

1½ 1/32 x 45°

3¼ 2½ 1¾ 1

5

Ø1.505/1.495

A

Ø1. DRILL, ⊥1¾

NOTES-
MACHINED SURFACES 125√ MIN.
BREAK ALL SHARP EDGES.

UNLESS OTHERWISE SPECIFIED DIMENSIONS ARE IN INCHES TOLERANCES ON	DRAWN BY JRW
	MACHINING SKILLS – LEVEL 1
FRACTIONS ± 1/64 DECIMALS ± 0.010 ANGLES ± 1°	DATE 8-12-XX
	TITLE 2.4 TURNING OPERATIONS – CHUCKING
	CHK'D MK
MATERIAL H.R. STEEL AISI 1018	HEAT TREATMENT NONE
	SCALE FULL 10
	5 OF 11
	DRAWING NO. B234793

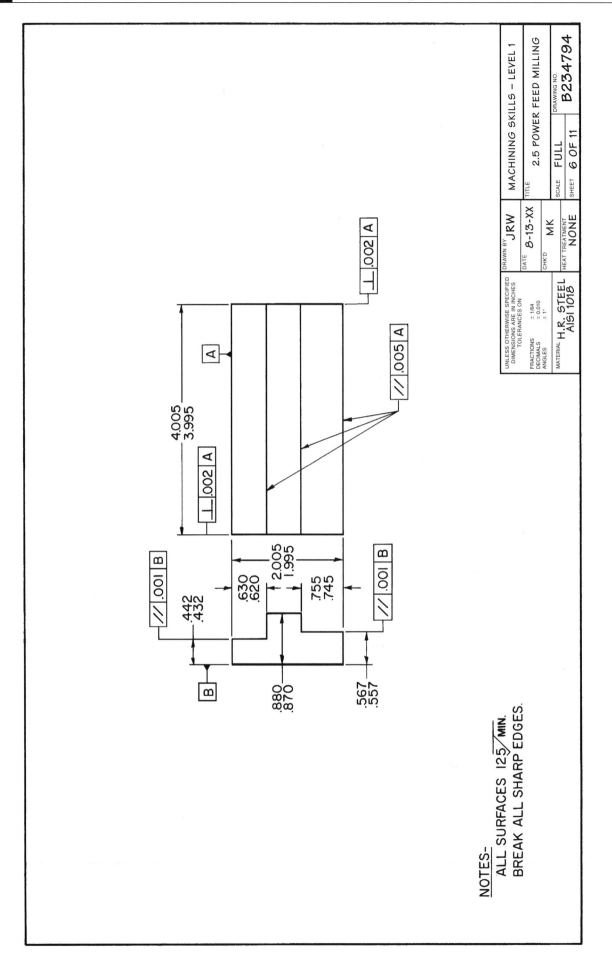

NOTES—
ALL SURFACES 125 MIN.
BREAK ALL SHARP EDGES.

| ⊥ | .002 | A |

| A |

| // | .005 | A |

4.005
3.995

| ⊥ | .002 | A |

| // | .001 | B |

.442
.432

.630
.620

2.005
1.995

.755
.745

.880
.870

.567
.557

| // | .001 | B |

| B |

UNLESS OTHERWISE SPECIFIED DIMENSIONS ARE IN INCHES TOLERANCES ON	DRAWN BY JRW	MACHINING SKILLS – LEVEL 1
FRACTIONS ± 1/64 DECIMALS ± 0.010 ANGLES ± 1°	DATE 8-13-XX	TITLE 2.5 POWER FEED MILLING
	CHK'D MK	
MATERIAL H.R. STEEL AISI 1018	HEAT TREATMENT NONE	SCALE FULL
		SHEET 6 OF 11
		DRAWING NO. B234794

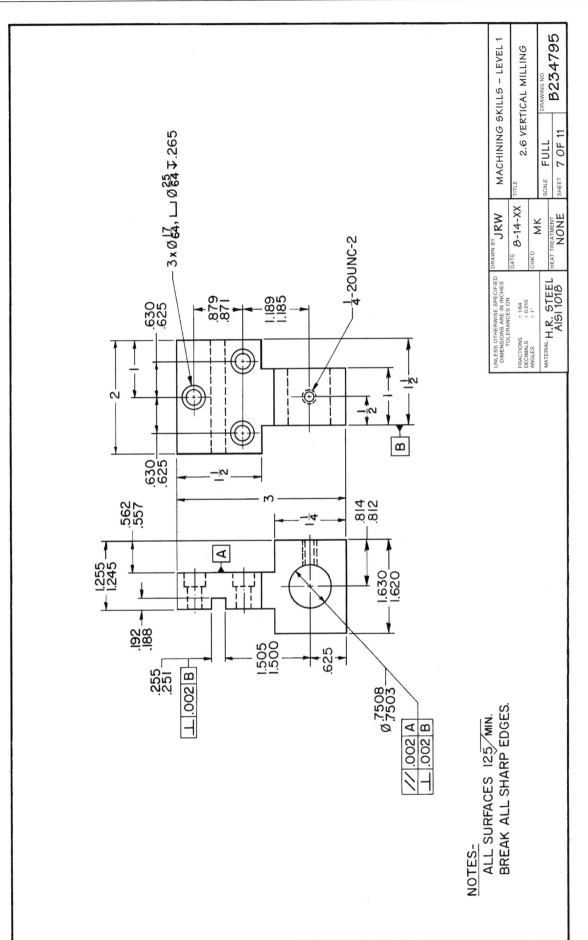

NOTES-
ALL SURFACES 125√ MIN.
BREAK ALL SHARP EDGES.

3 x Ø 17/64, ⌴ Ø 25/64 ⊼ .265

¼-20UNC-2

.630
.625

.879
.871

1.189
1.185

2

1

1½

1

½

B

.630
.625

1½

3

.562
.557

A

1¼

.814
.812

1.255
1.245

1.630
1.620

.192
.188

1.505
1.500

.625

.255
.251

⊥ .002 B

Ø .7508
.7503

∥ .002 A
⊥ .002 B

UNLESS OTHERWISE SPECIFIED
DIMENSIONS ARE IN INCHES
TOLERANCES ON

FRACTIONS ± 1/64
DECIMALS ± 0.010
ANGLES ± 1°

MATERIAL H.R. STEEL
AISI 1018

DRAWN BY JRW
DATE 8-14-XX
CHKD MK
HEAT TREATMENT NONE

MACHINING SKILLS – LEVEL 1
TITLE 2.6 VERTICAL MILLING
SCALE FULL
SHEET 7 OF 11

DRAWING NO.
B234795

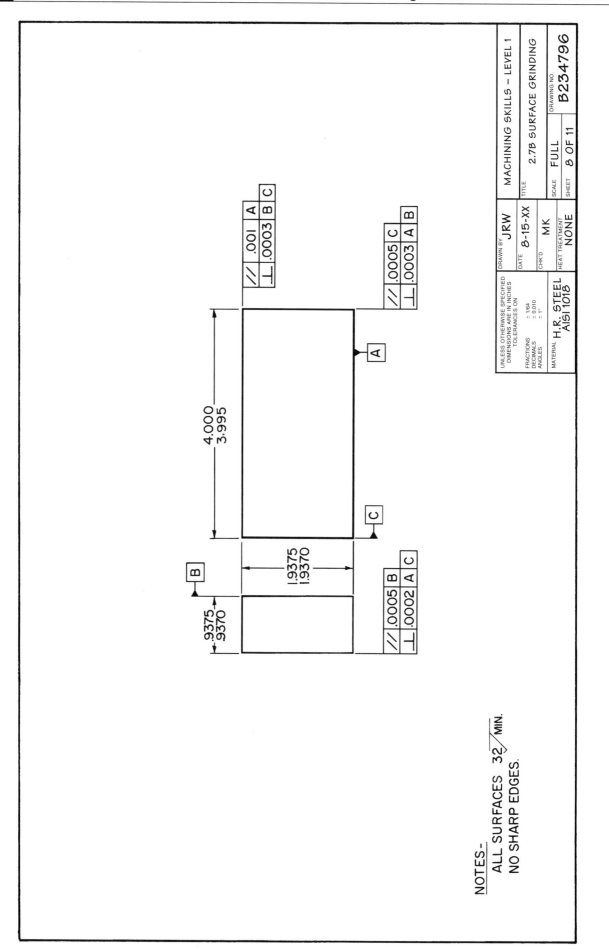

NOTES-
ALL SURFACES 32 / MIN.
NO SHARP EDGES.

//	.001	A	
⊥	.0003	B	C

//	.0005	C	
⊥	.0003	A	B

//	.0005	B	
⊥	.0002	A	C

4.000
3.995

1.9375
1.9370

.9375
.9370

UNLESS OTHERWISE SPECIFIED DIMENSIONS ARE IN INCHES TOLERANCES ON	DRAWN BY JRW	MACHINING SKILLS – LEVEL 1
FRACTIONS ± 1/64 DECIMALS ± 0.010 ANGLES ± 1°	DATE 8-15-XX	TITLE 2.7B SURFACE GRINDING
	CHK'D MK	
MATERIAL H.R. STEEL AISI 1018	HEAT TREATMENT NONE	SCALE FULL DRAWING NO. B234796
		SHEET 8 OF 11

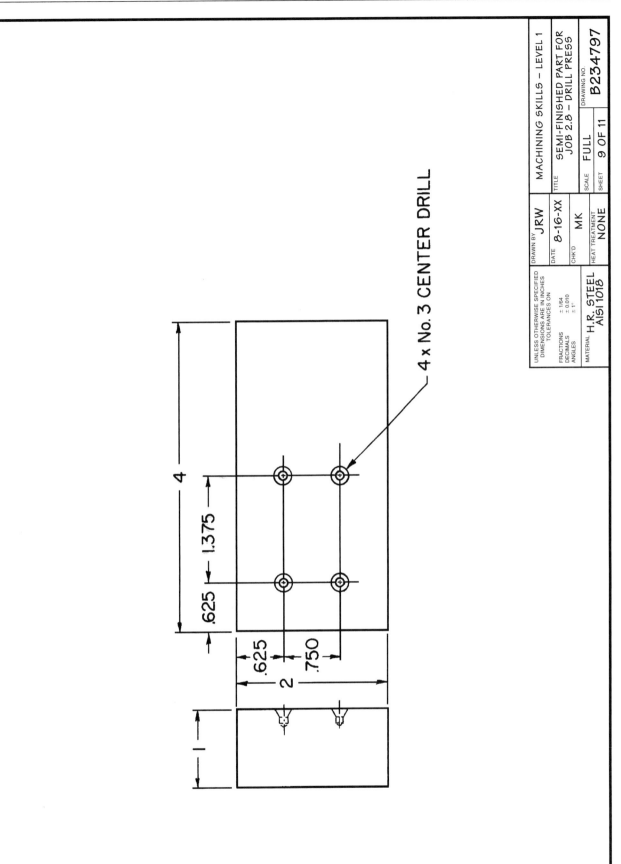

4 x No. 3 CENTER DRILL

DRAWN BY	JRW	
DATE	8-16-XX	
CHK'D	MK	
HEAT TREATMENT	NONE	

MACHINING SKILLS – LEVEL 1

TITLE: SEMI-FINISHED PART FOR JOB 2.8 – DRILL PRESS

DRAWING NO. B234797

SCALE: FULL SHEET: 9 OF 11

UNLESS OTHERWISE SPECIFIED
DIMENSIONS ARE IN INCHES
TOLERANCES ON

FRACTIONS ± 1/64
DECIMALS ± 0.010
ANGLES ± 1°

MATERIAL: H.R. STEEL AISI 1018

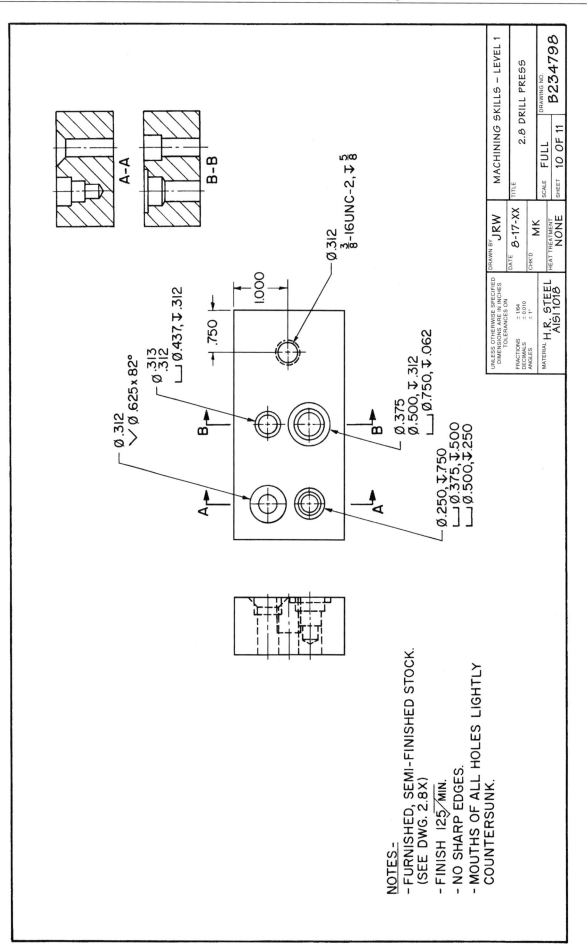

NOTES-
- FURNISHED, SEMI-FINISHED STOCK.
  (SEE DWG. 2.8X)
- FINISH 125 MIN.
- NO SHARP EDGES.
- MOUTHS OF ALL HOLES LIGHTLY
  COUNTERSUNK.

A-A

B-B

Ø.312
⅜-16UNC-2, ⌁ ⅝

Ø.437, ⌁.312

1.000

.750

Ø.313
Ø.312

Ø.375
Ø.500, ⌁.312
Ø.750, ⌁.062

Ø.312
Ø.625 x 82°

Ø.250, ⌁.750
Ø.375, ⌁.500
Ø.500, ⌁.250

UNLESS OTHERWISE SPECIFIED DIMENSIONS ARE IN INCHES TOLERANCES ON	DRAWN BY JRW	MACHINING SKILLS – LEVEL 1	
FRACTIONS ± 1/64 DECIMALS ± 0.010 ANGLES ± 1°	DATE 8-17-XX	TITLE 2.8 DRILL PRESS	
	CHK'D MK	SCALE FULL	DRAWING NO. B234798
MATERIAL H.R. STEEL AISI 1018	HEAT TREATMENT NONE	SHEET 10 OF 11	

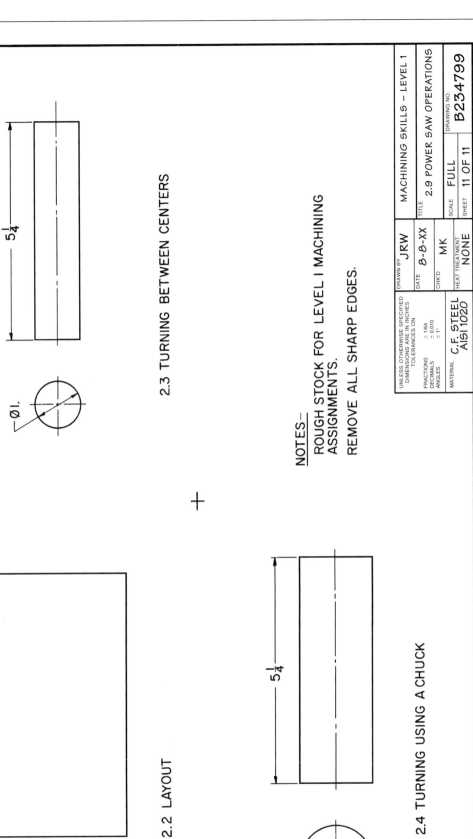

5 1/4

Ø1.

2.3 TURNING BETWEEN CENTERS

NOTES—
ROUGH STOCK FOR LEVEL I MACHINING ASSIGNMENTS.
REMOVE ALL SHARP EDGES.

6 1/4

3/8

3 1/4

2.2 LAYOUT

5 1/4

Ø1 3/4

2.4 TURNING USING A CHUCK

DRAWN BY JRW	MACHINING SKILLS – LEVEL 1
DATE 8-8-XX	TITLE 2.9 POWER SAW OPERATIONS
CHK'D MK	DRAWING NO. B234799
HEAT TREATMENT NONE	SCALE FULL
	SHEET 11 OF 11

UNLESS OTHERWISE SPECIFIED DIMENSIONS ARE IN INCHES
TOLERANCES ON
FRACTIONS ± 1/64
DECIMALS ± 0.010
ANGLES ± 1°

MATERIAL C.F. STEEL AISI 1020

Chapter 1

An Introduction to Machining Technology

LEARNING OBJECTIVES

After studying this chapter, students will be able to:
- ○ Discuss how modern machine technology affects the workforce.
- ○ Give a brief explanation of the evolution of machine tools.
- ○ Provide an overview of machining processes.
- ○ Explain how CNC machining equipment operates.
- ○ Describe the role of the machinist.

INSTRUCTIONAL MATERIALS

Text: pages 11–22
    Test Your Knowledge Questions, page 21
Workbook: pages 7–10
Instructor's Resource: pages 39–48
    Guide for Lesson Planning
    Research and Development Ideas
    Reproducible Masters:
        1-1 Bow Drill
        1-2 Bow Lathe
        1-3 Spring Pole Lathe
        1-4 Treadle Lathe
        1-5 Great Wheel Lathe
        1-6 Thread Cutter
        1-7 Test Your Knowledge Questions
    Color Transparencies (Binder/CD only)

GUIDE FOR LESSON PLANNING

The purpose of this chapter is to let your students know that the seemingly wondrous machines of today did not just happen. Their invention and development occurred over many centuries starting with the "Adam and Eve" of machine tools, the bow drill and bow lathe.

The construction of new machine tools or the improvement of existing tools continually makes the creation and development of new inventions possible. Machine tools began when humans wanted to extend the power of their hands over stubborn materials and devised the first bow drill. No one is sure where or when this first happened. Have students read and study Chapter 1, *Introduction to Machining Technology*. Review the assignment with them and discuss the following:

- The difficulty in naming a product that does not require the use of a machine tool somewhere in its manufacture or production.
- How could machine tools be made when there were no machine tools to make them?
- The evolution of the machine tool. Use the Reproducible Masters to show basic improvements in early machines.
- The importance of a standard system of measurement.
- Basic machine tool operation.
- The development of nontraditional machining processes.
- Automating the machining process.

Technical Terms

Review the terms introduced in the chapter. New terms can be assigned as a quiz, homework, or extra credit. These terms are also listed at the beginning of the chapter.

band machining
computer numerical control (CNC)
drill press
lathe
machine tools
machinist
milling machine
numerical control (NC)
precision grinding
skill standards

Review Questions

Assign *Test Your Knowledge* questions. Copy and distribute Reproducible Master 1-7 or have students use the questions on page 21 in the text and write their answers on a separate sheet of paper.

Workbook Assignment

Assign Chapter 1 of the *Machining Fundamentals Workbook.*

Research and Development

Discuss the following topics in class or have students complete projects on their own.
1. What was the Industrial Revolution?
2. Who invented the turret lathe?
3. Have students select one of the seven power sources listed in section 1.1.2 and write a brief history. Students should include drawings or photos of past and current uses.
4. The invention of the lathe is lost in history. The first lathe was believed to have been a "tree" lathe. Work was mounted between two trees and a limber branch fitted with a section of cord provided the power. Prepare a paper on the known history of the lathe. Illustrate it with transparencies designed for the overhead projector.

TEST YOUR KNOWLEDGE ANSWERS, PAGE 21

1. e. All of the above.
2. steam engine
3. In order: hand power, foot power, animal power, water power, steam power, central electrical power, and individual electrical power.
4. lathe
5. b. College graduates.
6. c. There was no standard of measurement.
7. The United States adopted a standard measuring system.
8. Evaluate individually. Refer to Section 1.3 of the text.
9. computer numerical control (CNC)
10. Student answers will vary but may include the following: permit machines to operate unattended; operate in hazardous and harsh environments; perform tedious operations; handle heavy materials; position parts with great repetitive precision.

ANSWERS TO THE WORKBOOK, Pages 7–10

1. reciprocating
2. d. All of the above.
3. b. *granddaddy* of all modern chip-making machine tools
4. Industrial Revolution
5. a. devised a system for mass production in the 1820s
6. Any order: hand power, foot power, animal power, water power, steam power, central electrical power, individual electrical power.
7. a. spiral flutes pulling the tool into the work
8. Any three of the following: electrical discharge machining (EDM), electrochemical machining (ECM), chemical milling, chemical blanking, hydrodynamic machining (HDM), ultrasonic machining, Electron beam machining (EBM), Laser machining, hexapods.
9. binary number
10. microchip
11. d. All of the above.
12. robotic
13. Student answers will vary but may include three of the following: make a thorough study of the print; determine the machining that must be done; ascertain tolerance requirements; plan the machining sequence; determine how the setup will be made; select the machine tool, cutter(s), and other tools and equipment that will be needed; calculate cutting speeds and feeds; select the proper cutting fluid for the material being machined.
14. skill standards
15. Evaluate individually.

Bow Drill

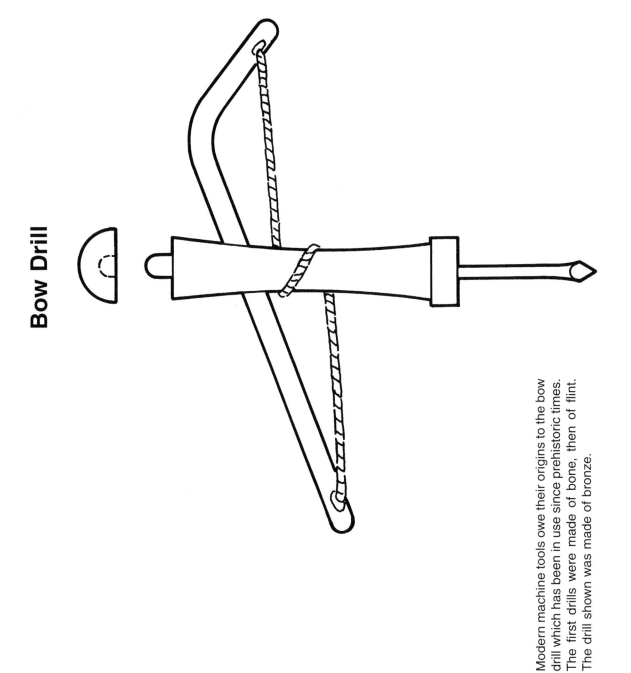

Modern machine tools owe their origins to the bow drill which has been in use since prehistoric times. The first drills were made of bone, then of flint. The drill shown was made of bronze.

1-2

Bow Lathe

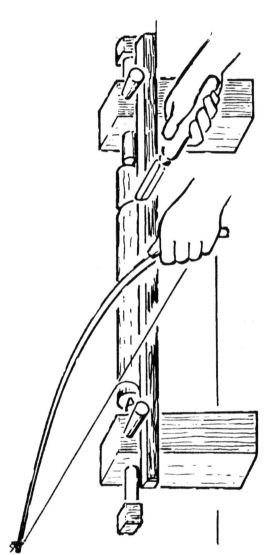

The bow lathe was developed from the bow drill and like the bow drill is used even today in India, the Far East, and some areas in Africa.

Spring Pole Lathe

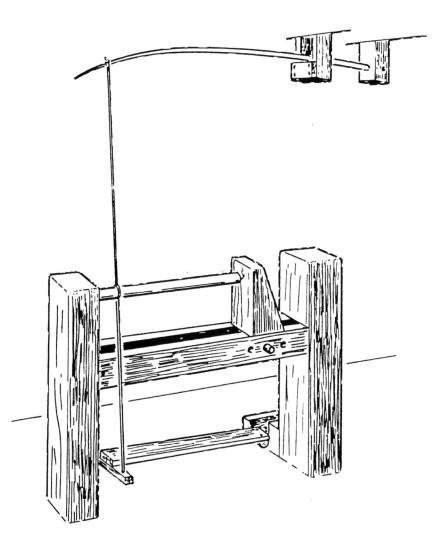

The spring pole lathe is thought to have been the next development in the evolution of machine tools. It was operated by the artisan's foot and left both hands free to handle the cutting tool.

1-3

Treadle Lathe

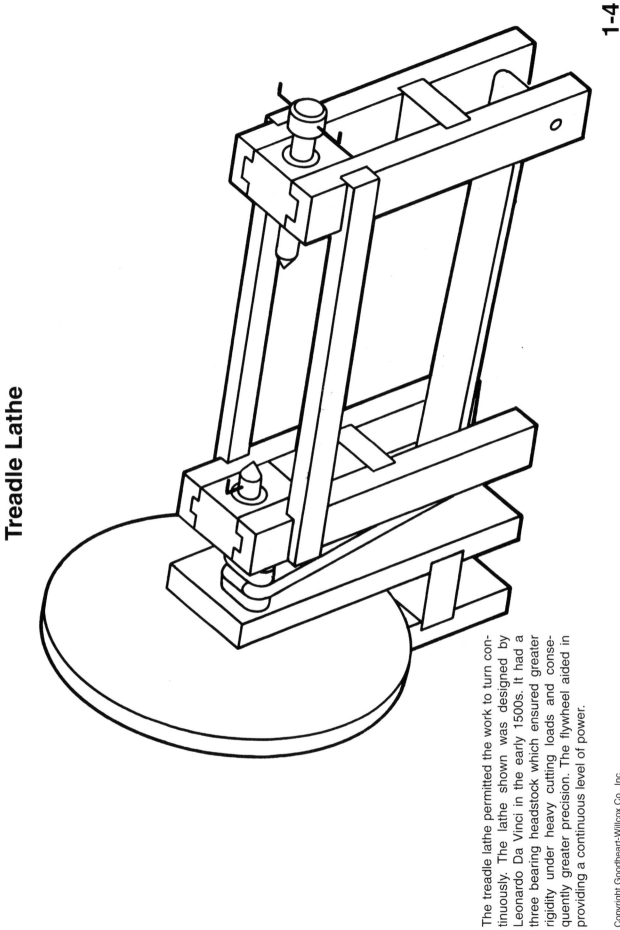

The treadle lathe permitted the work to turn continuously. The lathe shown was designed by Leonardo Da Vinci in the early 1500s. It had a three bearing headstock which ensured greater rigidity under heavy cutting loads and consequently greater precision. The flywheel aided in providing a continuous level of power.

Great Wheel Lathe

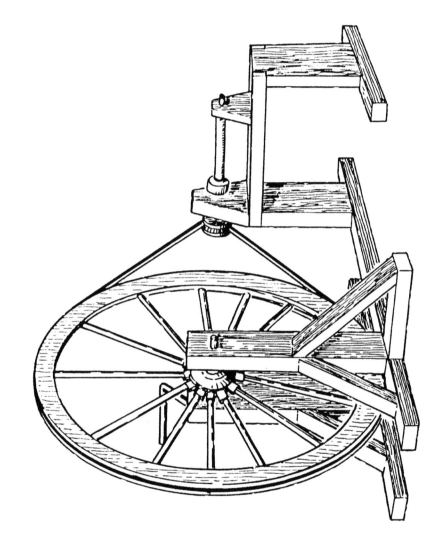

The great wheel lathe permitted skilled workers to keep all of their attention on the work. While the artisan worked at the lathe, another worker turned the wheel by hand.

Thread Cutter

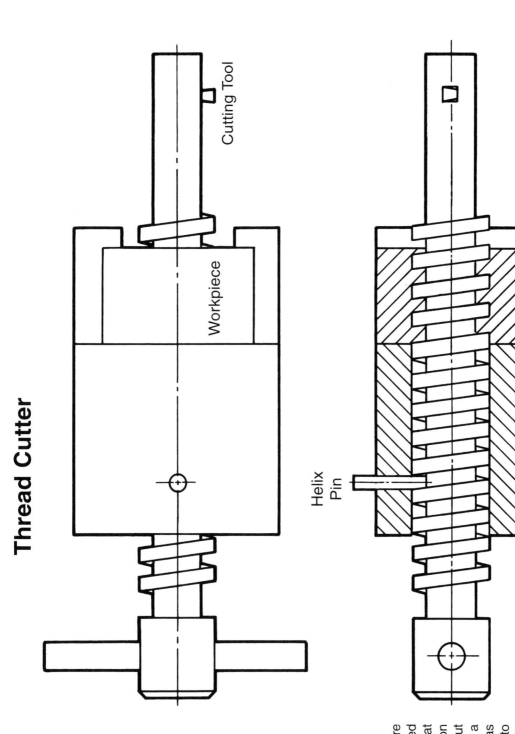

About AD 100, screw threads were made as part of screw presses used for pressing grapes and olives. At that time, the male threads could be cut on hard wood by hand. Making the nut was another matter. The device had a single-point cutting tool which was rotated through the workpiece (nut) to cut the threads shown. Later, a modified form of this device was used to cut threads in metal which was used as a die to make threads on longer metal rods.

An Introduction to Machining Technology

Name: _____ Date: _____ Score: _____

1. One of the first machine tools, the bow lathe, _____.   1. _____
   a. Could only turn softer materials
   b. Has been dated back to about 1200 BC
   c. Eventually gave way to treadle power
   d. None of the above.
   e. All of the above.

2. The Industrial Revolution could not have taken place   2. _____
   without the cheap convenient power of the _____.

3. List seven power sources in the order that they have evolved over the last 150 years or so.

   _____

   _____

   _____

   _____

   _____

   _____

   _____

4. Almost all machine tools have evolved from the _____.   4. _____

5. Jobs such as tool-and-diemaking and precision machining   5. _____
   require aptitudes comparable to those of _____.
   a. High school graduates.
   b. College graduates.
   c. High school equivalency graduates.
   d. All of the above.
   e. None of the above.

6. Eli Whitney's system of mass-production systems for   6. _____
   muskets had a major problem because _____.

   a. there were no skilled workers.
   b. there was no good source of power.
   c. there was no standard of measurement.
   d. All of the above.
   e. None of the above.

7. What occurred in the mid-1860s that was very important to the development of machining
   technology in the United States? _____

   _____

1-7

(continued)

Name: _____

8. List four types of nontraditional machining processes and briefly describe their operation.

_____

_____

_____

_____

_____

_____

9. The introduction of the microchip in the mid-1970s led to   9. _____
   the introduction of _____ machine tools.

10. List four industrial applications of robots.

Chapter 2

_____

_____

_____

_____

Shop Safety

LEARNING OBJECTIVES

After studying this chapter, students will be able to:
- ○ Give reasons why shop safety is important.
- ○ Explain why it is important to develop safe work habits.
- ○ Recognize and correct unsafe work practices.
- ○ Apply safe work practices when employed in a machine shop.
- ○ Select the appropriate fire extinguisher for a particular type of fire.

INSTRUCTIONAL MATERIALS

Text: pages 23–28
   Test Your Knowledge Questions, page 28
Workbook: pages 11–14
Instructor's Resource: pages 49–56
   Guide for Lesson Planning
   Research and Development Ideas
   Reproducible Masters:
      2-1 Fire Extinguishers and Fire
         Classifications
      2-2 Test Your Knowledge Questions
   Color Transparency (Binder/CD only)

GUIDE FOR LESSON PLANNING

Introduce the chapter with a display of safety equipment and safety posters. Discuss the duties of a safety officer in industry. Safety *cannot* be overemphasized!

Have students read and study the chapter. Review the assignment and discuss the following:
- State and shop/lab safety requirements.
- Students must assume responsibility for their safety and others in the shop/lab.
- Approved eye protection must be worn while working in the shop/lab.
- No machines to be operated until instructions have been given in its operation.
- No machines are to be operated unless all guards and safety devices are in place and functioning properly.
- Permission must be received before operating a machine tool.
- Dress must be appropriate.
- Students must avoid operating machine tools and other equipment while their senses are impaired by medication or other substances.
- Safe technique for handling metal chips and cuttings produced while operating machine tools.
- Importance of washing hands thoroughly after working in the machine shop/lab.
- Safe disposal of cloths used to clean machines.
- Procedure to be followed for reporting and taking care of any cut, burn, bruise, scratch, or puncture, no matter how minor it may appear.

Technical Terms

Review the terms introduced in the chapter. New terms can be assigned as a quiz, homework, or extra credit. These terms are also listed at the beginning of the chapter.

adequate ventilation
approved-type respirator
back injuries
combustible materials
electrocution
OSHA
protective clothing
safety equipment
unsafe practice
warning signs

Review Questions

Assign *Test Your Knowledge* questions. Copy and distribute Reproducible Master 2-2 or have students use the questions on page 28 in the text and write their answers on a separate sheet of paper.

Workbook Assignments

Assign Chapter 2 of the *Machining Fundamentals Workbook.*

Research and Development

Discuss the following topics in class or allow students to choose topics for individual or group projects.

1. Invite a local industry safety officer to discuss typical safety programs, and why industry places so much emphasis on a safe working environment.

2. Ask a representative from your local fire department to demonstrate the proper use of fire extinguishers and what to do in case of a fire in the shop. Copy and distribute Reproducible Master 2-1.

3. Invite a safety expert from a local shop or a safety equipment supplier to evaluate your safety program and make recommendations for any necessary improvements.

4. Work with school officials to see what protective eye wear is available for each student. Prepare a table of estimated needs, cost per year, and cost in succeeding years.

5. Develop and produce a series of colorful safety posters for the shop.

6. Prepare visual aids on safe work habits to be observed when using hand and machine tools.

7. Secure samples of safety posters and other safety program features used by local industry.

8. Design and construct a bulletin board on eye safety.

9. Contact the National Society for the Prevention of Blindness, 1790 Broadway, New York, New York 10019, for information on its Wise Owl Club of America.

10. Show a video or film on eye safety.

11. Produce a slide show or video on safe work habits when using the lathe, drill press, grinder, or vertical milling machine.

TEST YOUR KNOWLEDGE ANSWERS, Page 28

1. Evaluate individually.
2. carelessness
3. c. the entire time you are in the shop.
4. spontaneous combustion
5. Flying chips can cause serious eye injuries and vaporized oil can cause a fire and result in painful burns and property damage.
6. you have received instructions on its safe operation; all guards are in place; it has been determined that it is in safe operating condition
7. adjustments, measurements
8. brush, hands
9. adequate ventilation
   approved dust/fume mask
10. attention
11. heavy machine accessories or large pieces of metal stock
12. Check with your doctor, pharmacist, or school clinic to determine whether it will be safe for you to operate a machine.
13. Evaluate individually.

WORKBOOK ANSWERS, Pages 11–14

1. a. at all times when in the shop
2. habit
3. c. Both a and b.
4. brush, pliers
5. c. spontaneous combustion may result

6. d. Both b and c.
7. ventilation, dust mask
8. Evaluate individually.
9. d. All of the above.
10. e. Both a and c.
11. Any two of the following: when your senses are affected by medication, all guards and safety features are not in place, you have not been instructed in the safe operation of the machine.
12. a. When working in a noisy area.
    b. When handling solvents, cutting fluids, and oils.
    c. When machining produces airborne dust particles.

13. d. All of the above.
14. c. Class C Fires
15. b. Class B Fires
16. a. Class A Fires
17. d. Class D Fires
18. Evaluate individually.
19. Evaluate individually.
20. Evaluate individually.

Fire Extinguishers and Fire Classifications

Fires	Type	Use	Operation
Class A Fires Ordinary Combustibles (Materials such as wood, paper, textiles.) *Requires. . . cooling-quenching* **A**	**Soda-acid** Bicarbonate of soda solution and sulfuric acid	Okay for use on **A** Not for use on **B** **C** **D**	Direct stream at base of flame.
Class B Fires Flammable Liquids (Liquids such as grease, gasoline, oils, and paints.) *Requires. . . blanketing or smothering* **B**	**Pressurized Water** Water under pressure	Okay for use on **A** Not for use on **B** **C** **D**	Direct stream at base of flame.
	Carbon Dioxide (CO_2) Carbon dioxide (CO_2) gas under pressure	Okay for use on **B** **C** Not for use on **A** **D**	Direct discharge as close to fire as possible, first at edge of flames and gradually forward and upward.
Class C Fires Electrical Equipment (Motors, switches, and so forth.) *Requires. . . a nonconducting agent.* **C**	**Foam** Solution of aluminum sulfate and bicarbonate of soda	Okay for use on **A** **B** Not for use on **C** **D**	Direct stream into the burning material or liquid. Allow foam to fall lightly on tire.
Class D Fires Combustible Metals (Flammable metals such as magnesuim and lithium.) *Requires. . . blanketing or smothering.* **D**	**Dry Chemical**	Multi purpose type / Ordinary BC type Okay for: **A** **B** **C** / Okay for: **B** **C** Not okay for: **D** / Not okay for: **A** **D**	Direct stream at base of flames. Use rapid left-to-right motion toward flames.
	Dry Chemical *Granular type material*	Okay for use on **D** Not for use on **A** **B** **C**	Smother flames by scooping granular material from bucket onto burning metal.

2-1

Shop Safety

Name: _____ Date: _____ Score: _____

1. Why is shop safety so important? _____

   _____

   _____

   _____

   _____

   _____

2. Most shop accidents are caused by _____.   2. _____

3. Safety glasses should be worn _____.   3. _____
   a. most of the time
   b. only when working on machines
   c. the entire time you are in the shop
   d. None of the above.

4. Oily rags should be placed in a safety container to prevent   4. _____
   _____.

5. Why should compressed air not be used to clean chips from machine tools?

   _____

   _____

6. Never attempt to operate a machine until _____

   _____

   _____.

7. Always stop machine tools before making _____ and   7. _____
   _____.
                                                  _____

8. Use a(n) _____ to remove chips and shavings, *not* your   8. _____
   _____.
                                                  _____

9. When working in an area contaminated with dust or sol-   9. _____
   vent fumes, be sure there is _____. A(n) _____ should
   also be worn when working in a dusty area.   _____

10. Secure prompt _____ for any cut, bruise, scratch, or burn.   10. _____

11. Get help when moving _____

   _____.

Name: _____

12. What should you do before operating a machine tool if you are taking medication of any sort?

_____

_____

_____

13. Why is it necessary to take special precautions when handling long sections of metal stock?

Chapter 3

_____

_____

_____

_____

Understanding Drawings

LEARNING OBJECTIVES

After studying this chapter, students will be able to:
- ○ Read drawings that are dimensioned in fractional inches, decimal inches, and in metric units.
- ○ Explain the information found on a typical drawing.
- ○ Describe how detail, subassembly, and assembly drawings differ.
- ○ Point out why drawings are numbered.
- ○ Explain the basics of geometric dimensioning and tolerancing.

INSTRUCTIONAL MATERIALS

Text: pages 29–54
  Test Your Knowledge Questions,
  pages 52–53
Workbook: pages 15–22
Instructor's Resource: pages 57–70
  Guide for Lesson Planning
  Research and Development Ideas
  Reproducible Masters:
    3-1 Typical Assembly Drawing
    3-2 Alphabet of Lines
    3-3 Threads (*how depicted on a drawing*)
    3-4 Information on a Typical Drawing
    3-5 Dual Dimensioning
    3-6 Metric Drawing
    3-7 Geometric Dimensioning and
       Tolerancing (*application of*)
    3-8 Test Your Knowledge Questions
  Color Transparencies (Binder/CD only)

GUIDE FOR LESSON PLANNING

This chapter introduces and explains the basics of drawings used in industry. Since it would not be possible to manufacture complex products without them, the machinist must know how to obtain and understand all of the information provided on drawings.

With the increasing use of computer-generated machining programs, drawings may not always be available to the machinist at the work station. Often times, the machinist only sees a "drawing" on the computer monitor and the computer program makes corrections and adjustments or alerts the machinist to possible problems. However, the machinist may have to refer to the drawings to determine what adjustments and changes are acceptable. For this reason, it is of vital importance that a machinist be able to read and understand drawings.

Have students read and study Chapter 3. Review the assignment and discuss the following:
- • Importance of drawings to ensure that parts, no matter where they are made, will be interchangeable and fit properly in new assemblies and in similar assemblies made at an earlier date.
- • Reason for standardized symbols, lines, and figures.
- • The importance of the American National Standards Institute (ANSI).
- • The Alphabet of Lines.

- Symbols revised by ANSI and the symbols they replace.
- Information found on drawings and how it is used.
- Types of drawings used in shops.
- Methods used to reproduce drawings.
- Drawing sizes.
- Geometric Tolerancing and Dimensioning and why it is used.

Emphasize that a machinist:

- *Always* works to the dimensions, tolerances, and surface finishes specified on a drawing.
- *Never* scales a dimension from a drawing.

Technical Terms

Review the terms introduced in the chapter. New terms can be assigned as a quiz, homework, or extra credit. These terms are also listed at the beginning of the chapter.

> *actual size*
> *American National Standards Institute*
> *(ANSI)*
> *bill of materials*
> *dual dimensioning*
> *geometric dimensioning and tolerancing*
> *revisions*
> *scale drawings*
> *SI Metric*
> *US Conventional*
> *working drawings*

Review Questions

Assign *Test Your Knowledge* questions. Copy and distribute Reproducible Master 3-8 or have students use the questions on pages 52–53 in the text and write their answers on a separate sheet of paper.

Workbook Assignment

Assign Chapter 3 of the *Machining Fundamentals Workbook.*

Research and Development

Discuss the following topics in class or allow students to choose topics for individual or group projects.

1. Make a tracing and reproduce it using the diazo and electrostatic processes.
2. Prepare a display of the microfilming technique of print reproduction. Include prints, samples of film cards and photographs, or magazine advertisements illustrating the equipment used to make them.
3. Secure sample prints from industry.
4. Secure prints produced by the CAD (Computer-Aided Design) technique.
5. Prepare a display panel that shows a simple project from print to finished product.
6. Prepare transparencies for the overhead projector that show the title block, parts list, and material list from an actual industrial drawing. Use them to explain or describe an industrial drawing to the class. If possible, borrow a sample of the part shown on the drawing.
7. Contact a local industry and borrow prints of a simple assembly. If possible, also secure a sample of the object shown on the print. Develop a display.

TEST YOUR KNOWLEDGE ANSWERS, Pages 52–53

1. d. All of the above.
2. language of industry
3. one-millionth
4. one-millionth
5. Use a surface roughness comparison standard.
   Profilometer or electronic roughness gage.
6. bilateral
7. unilateral
8. b. Allowances in either oversize or under-size that a part can be made and still be acceptable.
9. scale drawings
10. a. Showing only a small portion of the complete object.
11. They sometimes get lost, damaged, or destroyed.
12. detail
13. It shows where and how the parts described on a detail drawing fit into the completed assembly.
14. Convenience in filing and locating drawings.
15. basic
16. reference
17. A feature control frame is used when a location or form tolerance is related to a datum.

18.

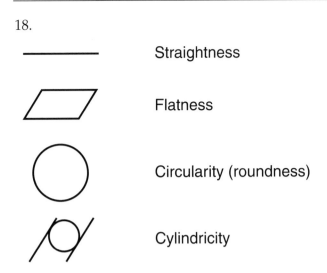

Straightness

Flatness

Circularity (roundness)

Cylindricity

19. Maximum material condition (MMC) is the condition in which the size of a feature contains the maximum amount of material within the stated limits of size. Also refer to Figure 3-31.

20. Least material condition (LMC) is the condition in which the size of a feature contains the least amount of material within the stated tolerance limits. Also refer to Figure 3-32.

WORKBOOK ANSWERS,
Pages 15–22

1. d. All of the above.
2. fractional
3. English, metric
4. Student answers will vary but may include any four of the following: material(s) to be used; surface finish required; tolerances; quantity of units per assembly; scale of drawing; next assembly or subassembly; revisions; the name of the object.
5. Tolerances are allowances, either undersize or oversize, permitted when machining or making an object.
6. roughness comparison
7. profilometer
8. d. Dimensions should *never* be scaled off a drawing.
9. b. only a small portion
10. drawings might be lost, damaged, or destroyed; same print may be needed in different places at same time
11. d. All of the above.
12. Evaluate individually.

13. When the amount of variation (tolerances) in form (shape and size) and position (location) needs to be more strictly defined, it provides the precision needed to allow for the most economical manufacture of parts.
14. Geometric dimensioning and tolerancing is a system that provides additional precision compared to conventional dimensioning. It ensures that parts can be easily interchanged.
15. They are employed to provide clarity and precision in communicating design specifications.
16. d. All of the above.
17. b. basic dimension
18. a. reference dimension
19. measured size of a part after it is manufactured
20. feature control frame
21. Maximum material (MMC)
22. Least material (LMC)
23. A. Material to be used
    B. Tolerances
    C. Quantity
    D. Scale
    E. Next assembly
    F. Revisions
    G. Name of object
    H. Drawing number
24. A. 3.000"
    B. 2.000"
    C. 1.625"
    D. 0.7503"
    E. +0.0003"
    F. 0.266"
    G. 0.265"
    H. 0.391"
    I. Remove burrs. Break sharp edges .010" Max. Finish 125/ all over except as noted.
25. A. Clamp, Alignment
    B. Full size
    C. B123456
    D. D45678
    E. Dual dimensioning
    F. Aluminum 6061-T4
    G. 12
    H. Distance from centerline to flat on top of part was changed from 1.50" (38.0 mm) to 1.62" (41.14 mm).

26. L. 4.25" (107.85 mm)
    W. 2.62" (66.54 mm)
    T. 0.75" (19.0 mm)
27. 1. 125/[3.2/] all over.
    2. Break all sharp edges 0.01 [0.5] MAX.
    3. Dimensions in [ ] are millimeters.
28. A. 1.50" (38.1 mm)
    B. 0.500"
29. No standard metric tool available this size.

30. A. 3.25" (82.45 mm)
    B. 1/4-20UNC-2
    C. There is no metric thread this size.
    D. 0.37" (9.5 mm)
    E. 0.26" (6.7 mm)
    F. 0.13" (3.5 mm)
    G. 0.75" (19.0 mm)
    H. +0.001"
    I. 0.75" (19.0 mm)
    J. 1.00" (25.4 mm)
    K. 1.62" (41.14 mm)

Typical Assembly Drawing

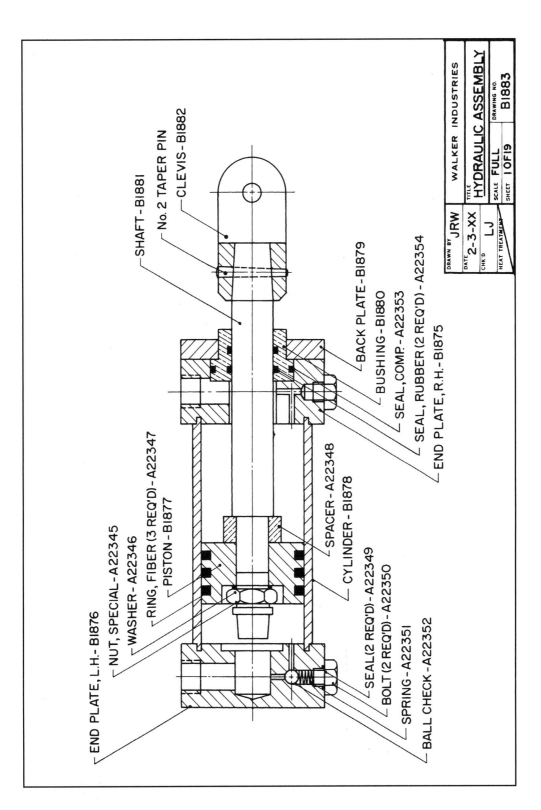

3-2

Alphabet of Lines

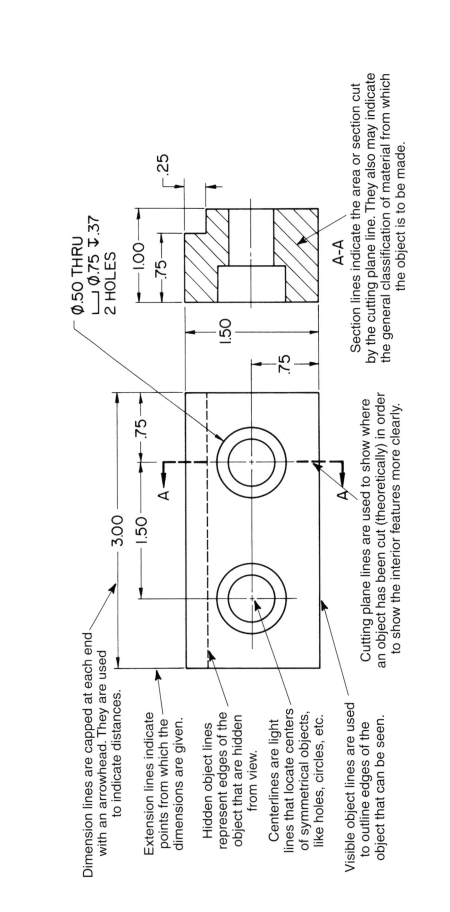

Ø.50 THRU
⌴Ø.75 ⌵.37
2 HOLES

A-A

Section lines indicate the area or section cut by the cutting plane line. They also may indicate the general classification of material from which the object is to be made.

Dimension lines are capped at each end with an arrowhead. They are used to indicate distances.

Extension lines indicate points from which the dimensions are given.

Hidden object lines represent edges of the object that are hidden from view.

Centerlines are light lines that locate centers of symmetrical objects, like holes, circles, etc.

Visible object lines are used to outline edges of the object that can be seen.

Cutting plane lines are used to show where an object has been cut (theoretically) in order to show the interior features more clearly.

Threads

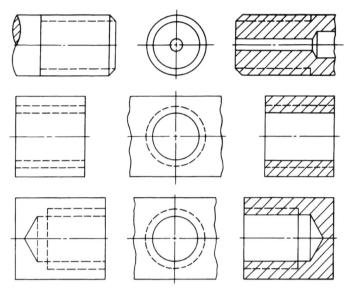

Simplified Representation

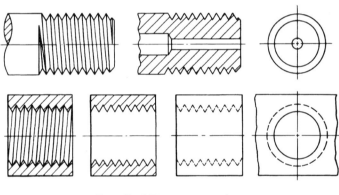

Detailed Representation

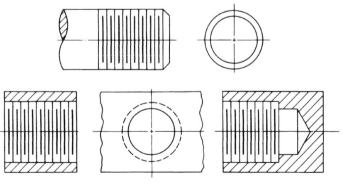

Schematic Representation

3-3

3-4

Information on a Typical Drawing

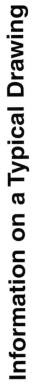

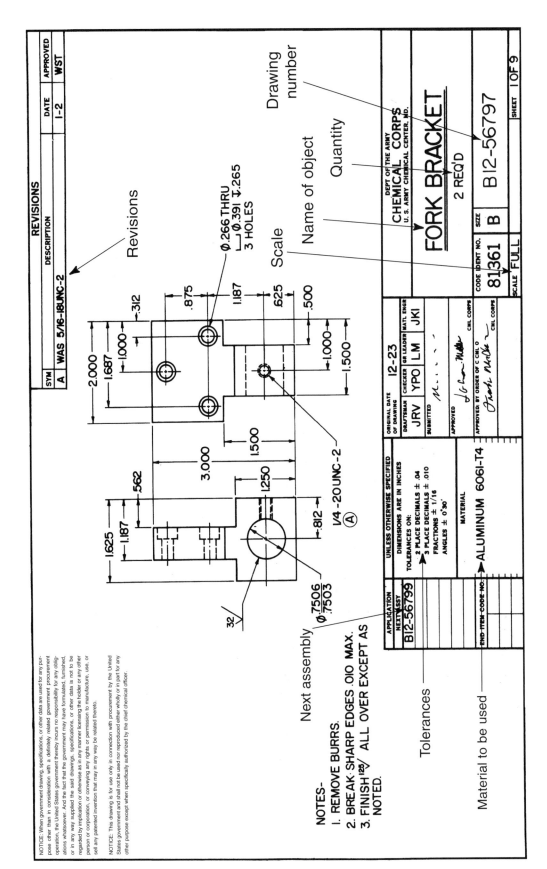

Dual Dimensioning

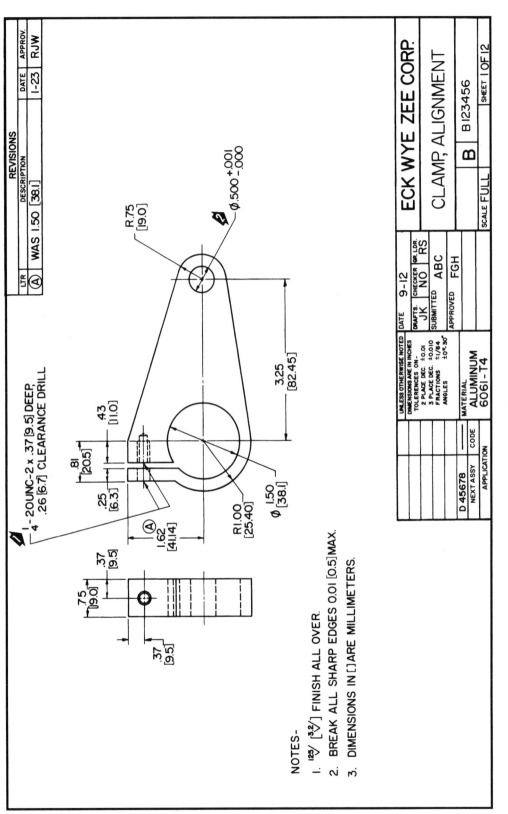

3-6

Metric Drawing

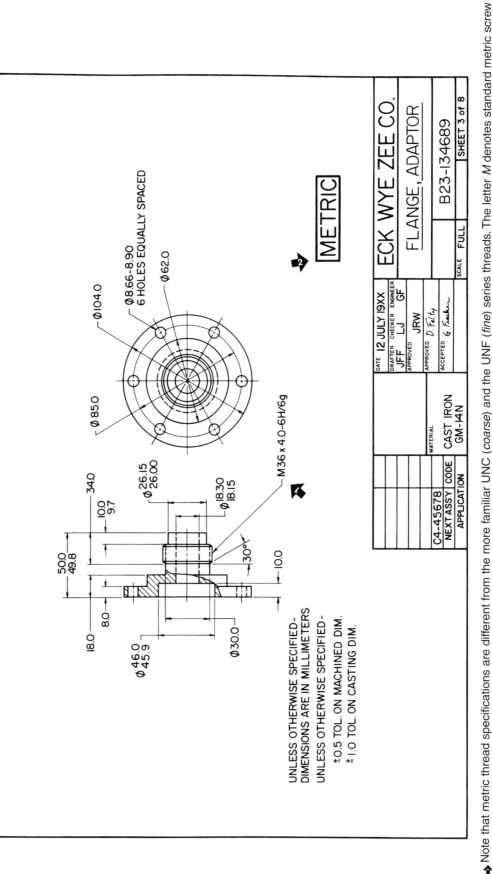

⌀104.0

⌀8.66–8.90
6 HOLES EQUALLY SPACED

⌀62.0

⌀85.0

M36 × 4.0–6H/6g

34.0

⌀26.15
26.00

10.0
9.7

⌀18.30
18.15

50.0
49.8

30°

10.0

18.0

8.0

⌀46.0
45.9

⌀30.0

METRIC

2

UNLESS OTHERWISE SPECIFIED–
DIMENSIONS ARE IN MILLIMETERS

UNLESS OTHERWISE SPECIFIED–
±0.5 TOL. ON MACHINED DIM.
±1.0 TOL. ON CASTING DIM.

			DATE 12 JULY 19XX	ECK WYE ZEE CO.
			DRAFTER JFF	CHECKER LJ	ENGINEER GF	FLANGE, ADAPTOR
			APPROVED JRW	
			APPROVED D. Felty	
	MATERIAL		ACCEPTED G Fischer	B23-I34689
	CAST IRON GM-I4N			SCALE FULL	SHEET 3 of 8
C4-45678				
NEXT ASSY	CODE			
APPLICATION				

⬆ Note that metric thread specifications are different from the more familiar UNC (*coarse*) and the UNF (*fine*) series threads. The letter *M* denotes standard metric screw threads. The 36 indicates the nominal thread diameter in millimeters. The 4.0 denotes thread pitch in millimeters. The 6H and 6g ate tolerance calss designations.

⬆ To avoid possible misunderstanding, *metric* is shown on the drawing in large letters.

Geometric Dimensioning and Tolerancing

A Geometric
characteristic
symbol

B Diameter symbol
(when used)
Zone descriptor

C Geometric tolerance

D Material condition
symbol

E Primary datum
reference

F Secondary
datum reference

G Tertiary datum
reference

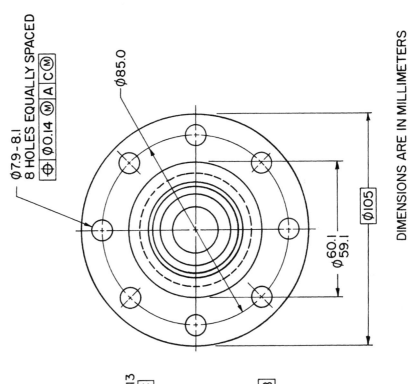

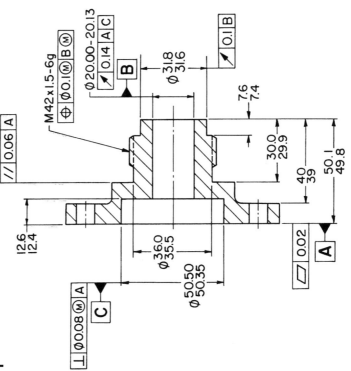

DIMENSIONS ARE IN MILLIMETERS

3-7

Understanding Drawings

Name: _____ Date: _____ Score: _____

1. Drawings are used to:
   a. Show, in multiview, what an object looks like before it is made.
   b. Standardize parts.
   c. Show what to make and the sizes to make it.
   d. All of the above.
   e. None of the above.

2. The symbols, lines, and figures that make up a drawing are frequently called the _____.

3. A microinch is _____ of an inch.

4. A micrometer is _____ of a meter.

5. How can surface roughness of a machined part be checked against specifications on the drawing?

   _____

   How can it be measured electronically? _____

   _____

   _____

6. When tolerances are plus and minus, it is called a _____ tolerance.

7. When tolerances are only plus or only minus, it is called a _____ tolerance.

8. Tolerances are:
   a. The different materials that can be used.
   b. Allowances in either oversize or undersize that a part can be made and still be acceptable.
   c. Dimensions.
   d. All of the above.
   e. None of the above.

9. Drawings made other than actual size are called _____.

10. A subassembly drawing differs from an assembly drawing by:
    a. Showing only a small portion of the complete object.
    b. Making it possible to use smaller drawings.
    c. Showing the object without all needed dimensions.
    d. All of the above.
    e. None of the above.

1. _____

2. _____

3. _____

4. _____

6. _____

7. _____

8. _____

9. _____

10. _____

3-8

(continued)

Name _____

11. Why are prints used in place of the original drawings? _____

_____

_____

12. The craft worker is given all of the information needed   12. _____
    to make a part on a _____ drawing.

13. What does an assembly drawing show? _____

_____

_____

14. Why are standard size drawing sheets used? _____

_____

_____

15. All dimensions have a tolerance except _____ dimensions.   15. _____

16. Dimensions placed between parentheses are _____   16. _____
    dimensions.

17. When is a feature control frame employed? _____

_____

_____

18. Sketch the form geometric tolerance symbols and indicate what they mean.

3-8

(continued)

Name _____

19. Define the term *maximum material condition* (MMC). Use a sketch if necessary. _____

_____

_____

_____

20 Define the term *least material condition* (LMC). Use a sketch if necessary. _____

_____

_____

_____

Chapter 4

Measurement

LEARNING OBJECTIVES

After studying this chapter, students will be able to:
○ Measure to 1/64″ (0.5 mm) with a steel rule.
○ Measure to 0.0001″ (0.002 mm) using a Vernier micrometer caliper.
○ Measure to 0.001″ (0.02 mm) using Vernier measuring tools.
○ Measure angles to 0°5′ using a universal Vernier bevel.
○ Identify and use various types of gages found in a machine shop.
○ Use a dial indicator.
○ Employ the various helper measuring tools found in a machine shop.

INSTRUCTIONAL MATERIALS

Text: pages 55–80
　　Test Your Knowledge Questions,
　　pages 78–80
Workbook: pages 23–32
Instructor's Resource: pages 71–92
　　Guide for Lesson Planning
　　Research and Development Ideas
　　Reproducible Masters:
　　　　4-1 Inch-Based Rule
　　　　4-2 Metric Rule
　　　　4-3 Inch-Based Micrometer
　　　　4-4 Inch-Based Vernier Micrometer
　　　　4-5 Metric Micrometer
　　　　4-6 Metric-Based Vernier Micrometer
　　　　4-7 25-Division Inch-Based Vernier
　　　　　　Caliper
　　　　4-8 25-Division Metric Vernier Caliper
　　　　4-9 50-Division Inch-Based Vernier
　　　　　　Caliper
　　　　4-10 50-Division Metric Vernier Caliper
　　　　4-11 Universal Bevel Protractor
　　　　4-12 Test Your Knowledge Questions
　　Color Transparencies (Binder/CD only)

GUIDE FOR LESSON PLANNING

Since this chapter is extensive, it is recommended that it be divided into four parts.

Part I—The Rule

The ability to make accurate measurements is basic to all types of skilled occupations. As a pretest, use the rules on Reproducible Masters 4-1 and 4-2 to determine a starting point for teaching measurement.

Have students read and study pages 55–57. Review the assignment and demonstrate and discuss the following:

• The various types of rules.

• How to read and use the various types of rules.

• How to make accurate measurements with a rule.

• How to handle and care for rules so they will retain their accuracy.

Part II—Micrometers

Once students become proficient with rules, introduce the micrometer caliper. Since most training centers have a limited number of the various types of micrometers, Reproducible Masters 4-3, 4-4, 4-5, and 4-7 have been provided for use as overhead transparencies or handouts.

Have students read and study pages 57–63. Review the assignment and demonstrate the following:

- The various types of micrometers.
- How to read inch- and metric-based micrometers.
- How to read inch- and metric-based Vernier micrometers.
- The proper way to use micrometers.
- The proper way to care for micrometers so they will retain their accuracy.

Allow students to examine and use micrometers. Provide various sizes of work so students can practice using and reading micrometers.

Part III—Vernier Measuring Tools

Have students read and study pages 63–67. Use Reproducible Masters 4-7, 4-8, 4-9, 4-10, and 4-11 to make overhead transparencies and handouts. Review the assignment and demonstrate the following:

- The various types of Vernier calipers.
- How to read inch- and metric-based Vernier calipers.
- How to read dial calipers.
- How to read the universal bevel protractor.
- The correct way to use Vernier and dial calipers.
- The proper way to care for Vernier measuring tools so they will retain their accuracy.

Allow students to examine and use the Vernier type tools. Provide various sizes of work so students can practice using and reading the tools.

Part IV—Gages and Dial Indicators

Have as many of the tools described in this section that are available in the shop on hand for student examination.

Have students read and study pages 67–78, paying particular attention to the illustrations. Discuss and demonstrate how the tools are used.

Technical Terms

Review the terms introduced in the chapter. New terms can be assigned as a quiz, homework, or extra credit. These terms are also listed at the beginning of the chapter.

dial indicators
gage blocks
gaging
graduations
helper measuring tools
International System of Units
metrology
micrometer caliper
steel rule
Vernier caliper

Review Questions

Assign *Test Your Knowledge* questions. Copy and distribute Reproducible Master 4-12 or have students use the questions on pages 78–80 in the text and write their answers on a separate sheet of paper.

Workbook Assignment

Assign Chapter 4 of the *Machining Fundamentals Workbook.*

Research and Development

Discuss the following topics in class or allow students to choose topics for individual or group projects.

1. Make a large working model of the hub and thimble of a micrometer. Use different size cardboard mailing tubes.
2. Develop a working model of a Vernier caliper. Make it large enough to be used to instruct the class. The model may be of a 25- or a 50-division scale.
3. Make an enlarged section of a No. 4 rule at least ten times actual size. Use basswood, plywood, or hardboard.
4. Prepare a transparency of a No. 4 rule using several overlays that can be used on an overhead projector to teach beginners how to read a rule.
5. Design and make a series of posters showing how to read a metric micrometer.
6. Prepare a display of various types of gages. Secure samples of work checked by gage type measuring tools.
7. Arrange for someone to demonstrate how optical flats are used. Use a film or video

presentation if the actual equipment cannot be secured.

8. Invite a quality control expert from a local industry to speak to the class or grant an interview. Ask them to describe their job and the specialized measuring tools that are used on the job. Prepare questions in advance that the class would like to have answered or explained.

9. Prepare a research paper on how temperature changes can affect measuring accuracy.

   - Prepare 1.000" (25.00 mm) long pieces of aluminum, brass, steel, plastic, and cast iron with exactly the same size cross section.
   - Record their exact lengths at room temperature with a Vernier micrometer caliper.
   - Place the sections in a freezer for 24 hours and quickly measure them again.
   - Record your findings. Place the sections in boiling water or in a heat treating furnace for 15 minutes at 200°F (93°C). Quickly measure them again. Record your findings.
   - Record a graph that will show how sizes varied under the three conditions of temperature. Using this information, have a class discussion how products can be affected by great changes in temperature and how industry takes into account this problem when certain products are designed.

10. Prepare a report or presentation on early measuring tools and some of the problems encountered before measuring standards were established.

TEST YOUR KNOWLEDGE ANSWERS, Pages 78–80

1.
1. 3/16	10. 1/32
2. 11/16	11. 13/32
3. 1 5/16	12. 21/32
4. 1 13/16	13. 25/32
5. 2 7/16	14. 1 3/32
6. 2 13/16	15. 1 15/32
7. 3 1/16	16. 1 25/32
8. 3 7/16	17. 2 5/32
9. 3 15/16	18. 2 17/32

19. 2 23/32	N. 0.50 mm
20. 2 29/32	O. 5.5 mm
21. 3 9/32	P. 11.5 mm
22. 3 7/16	Q. 19.5 mm
23. 3 19/32	R. 24.0 mm
24. 3 29/32	S. 32.5 mm
A. 300 mm	T. 36.5 mm
B. 295 mm	U. 43.5 mm
C. 289 mm	V. 54.5 mm
D. 284 mm	W. 62.5 mm
E. 278 mm	X. 74.5 mm
F. 273 mm	Y. 83.5 mm
G. 267 mm	Z. 88.5 mm
H. 262 mm	
I. 251 mm	
J. 241 mm	
K. 234 mm	
L. 227 mm	
M. 214 mm	

2.
A. 8.683"	F. 5.008"
B. 4.107"	G. 55.78 mm
C. 7.500"	H. 73.34 mm
D. 3.150"	I. 71.70 mm
E. 8.793"	J. 24.84 mm

3. mike

4. microinch

5. micrometer

6. 0.001, 0.0001, 0.01, 0.002

7. Student answers will vary but may include two of the following: it can be used to make both internal and external measurements; newer versions make both inch based and metric measurements; it has a larger measuring range.

8. 0.001, 0.02

9. Evaluate individually.

10. universal Vernier bevel protractor

11. Double end plug gage has the GO plug on one end and the NO-GO plug on the other end. The progressive or step plug gage has GO and NO-GO plugs on same end permitting gaging to be done in a single motion.

12. diameters, tolerance

13. Jo

14. a. Air pressure leakage between the plug and hole walls.

15. Continuous and balanced types.

16. Evaluate individually.
17. Optical flats
18. optical comparator
19. screw pitch gage
20. Evaluate individually. Refer to Figure 4-52.
21. Helper tools are not direct reading and require the help of a rule, micrometer, or Vernier caliper to determine the size of the measurement taken.
22. Compress the contact legs.

    Insert the gage into the hole and allow the legs to expand.

    After the proper fitting is obtained, lock the contacts into position.

    Remove the gage from the hole and make your reading with a micrometer.

23. A. 0.312″   G. 0.437″
    B. 0.625″   H. 0.937″
    C. 5.78 mm   I.   4.03 mm
    D. 0.375″   J.   0.500″
    E. 0.562″   K. 0.187″
    F. 9.67 mm   L. 16.07 mm

WORKBOOK ANSWERS,
Pages 23–32

1. A. 1/4   H. 2 7/32
   B. 7/8   I.   2 19/32
   C. 1 3/8   J.   2 25/32
   D. 2 1/8   K. 2 31/32
   E. 2 5/8   L. 3 11/32
   F. 3 1/2   M. 3 1/2
   G. 5/16   N. 3 21/32
   H. 11/16   O. 3 13/16
   I. 1 3/16   P.   3 31/32
   J. 1 11/16   Q. 5/64
   K. 2 7/16   R. 13/64
   L. 2 15/16   S. 23/64
   M. 3 9/16   T. 37/64
   N. 4 1/16   U. 45/64
2. A. 3/32   V. 55/64
   B. 15/32   W. 1 3/64
   C. 23/32   X. 1 10/64
   D. 27/32   Y. 1 19/64
   E. 1 5/32   Z. 1 27/64
   F. 1 17/32   AA. 1 39/64
   G. 1 27/32   BB. 1 61/64

      CC.   2 7/64
      DD.   2 15/64
      EE.   2 31/64
      FF.   2 39/64
      GG.   2 49/64
      HH.   2 63/64
      II.   2 9/64
      JJ.   3 17/64
      KK.   3 25/64
      LL.   3 33/64
      MM.   3 41/64
      NN.   3 58/64
      OO.   4 1/64

3. A. 305.0
   B. 294.0
   C. 286.0
   D. 281.0
   E. 272.0
   F. 266.0
   G. 261.0
   H. 255.0
   I. 249.0
   J. 241.0
   K. 233.0
   L. 225.0
   M. 214.0
   N. 6.5
   O. 12.5
   P. 20.5
   Q. 25.5
   R. 33.5
   S. 43.5
   T. 54.5
   U. 64.5
   V. 75.5
   W. 84.5
   X. 88.5
   Y. 93.5
   Z. 99.5
4. A. 0.856″
   B. 0.663″
5. A. 0.817″
   B. 0.532″
6. A. 0.748″
   B. 0.142″

7. A. 0.429″
   B. 0.081″
8. A. 0.357″
   B. 0.759″
9. A. 5.04 mm
   B. 12.99 mm
10. A. 1.39 mm
    B. 19.51 mm
11. A. 0.56 mm
    B. 14.61 mm
12. A. 9.62 mm
    B. 15.99 mm
13. A. 12.18 mm
    B. 13.83 mm
14. A. 0.743
    B. 4.157
15. A. 6.991
    B. 12.108
16. A. 8.475
    B. 11.708
17. A. 5.057
    B. 3.343
18. A. 75.34 mm
    B. 43.78 mm
19. A. 78.66 mm
    B. 23.66 mm
20. A. 69.28 mm
    B. 113.94 mm
21. d. Both b and c.
22. 0.001, 0.0001
23. 0.01 mm, 0.002 mm
24. Evaluate individually. Refer to Section 4.3.
25. Evaluate individually. Refer to Section 4.3.5.
26. 1/12, 5
27. go-no go
28. go-no go dimensions in one motion
29. air pressure leakage between the plug and diameter being measured
30. d. All of the above.
31. light waves
32. A measuring device that makes use of an enlarged image of the part to be inspected. The image is projected on a screen where it is superimposed on an enlarged accurate drawing of the part.
33. telescoping gage

34. the number of threads per inch or mm on a threaded section
35. The thin steel blades of a fillet and radius gage, are used to check concave and convex radii on corners or against shoulders. The gage is used for layout work and inspection, and as a template when grinding form cutting tools.
36. Measuring devices that are not direct reading but require the aid of a rule, micrometer, or Vernier caliper to determine size of measurement taken.
37. Inside caliper, telescoping gage, outside caliper, and small hole gage.
38. d. All of the above.
39. d. All of the above.
40. c. openings that are too small for a telescoping gage

ANSWERS FOR REPRODUCIBLE MASTERS

4-1 Inch-Based Rule

1. 1/16″	13. 9/32″
2. 1/4″	14. 1 5/32″
3. 1 1/8″	15. 1 25/32″
4. 1 13/16″	16. 2 11/32″
5. 7/16″	17. 2 29/32″
6. 2 15/16″	18. 3 19/32″
7. 2 11/16″	19. 4 7/32″
8. 4 3/16″	20. 4 17/32″
9. 4 3/4″	21. 5 1/32″
10. 5 5/8″	22. 5 9/32″
11. 5 7/8″	23. 5 21/32″
12. 3/32″	24. 5 31/32″

4-2 Metric Rule

1. 4.0 mm	11. 2.5 mm
2. 12 mm	12. 13.5 mm
3. 21 mm	13. 23.5 mm
4. 29 mm	14. 40.5 mm
5. 42 mm	15. 49.5 mm
6. 53 mm	16. 60.5 mm
7. 65 mm	17. 71.5 mm
8. 71 mm	18. 87 mm
9. 82 mm	19. 94.5 mm
10. 98 mm	20. 101.5 mm

Inch-Based Rule

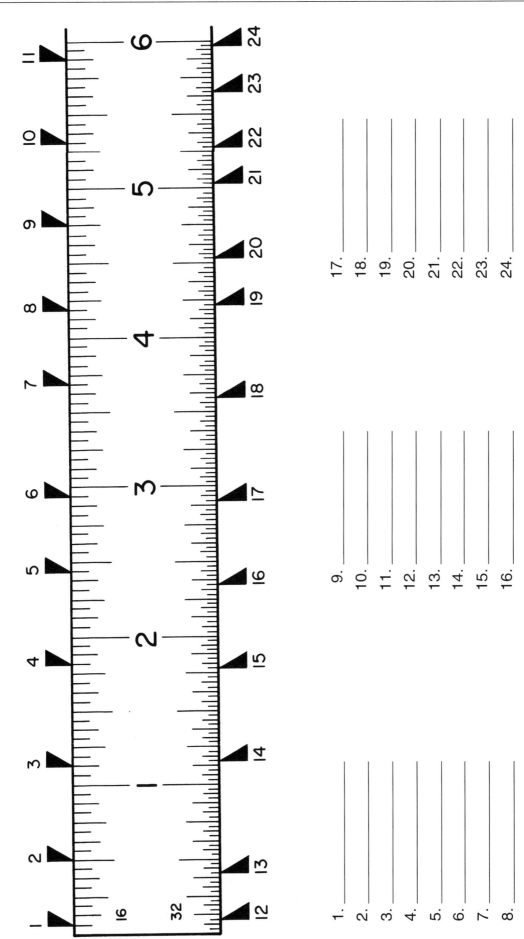

1. _____
2. _____
3. _____
4. _____
5. _____
6. _____
7. _____
8. _____

9. _____
10. _____
11. _____
12. _____
13. _____
14. _____
15. _____
16. _____

17. _____
18. _____
19. _____
20. _____
21. _____
22. _____
23. _____
24. _____

4-1

4-2

Metric Rule

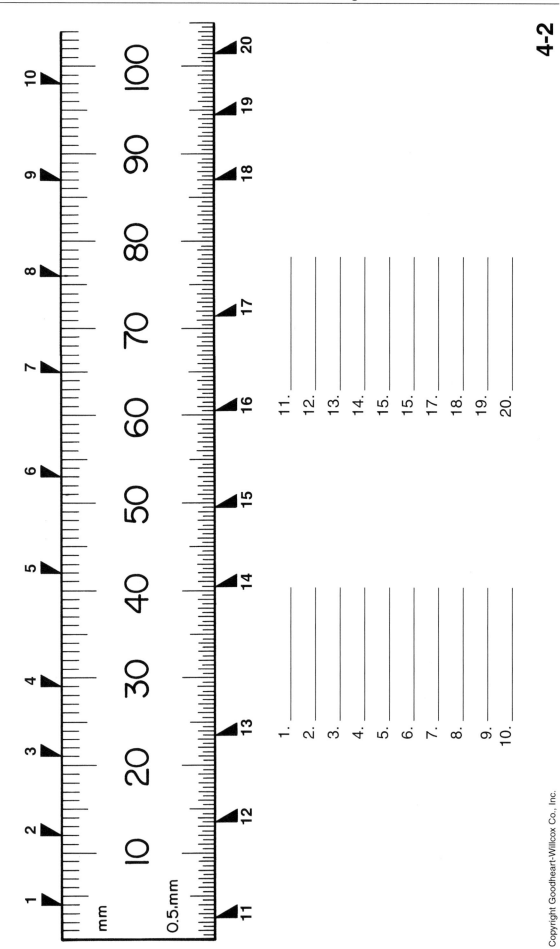

1. _____
2. _____
3. _____
4. _____
5. _____
6. _____
7. _____
8. _____
9. _____
10. _____

11. _____
12. _____
13. _____
14. _____
15. _____
15. _____
17. _____
18. _____
19. _____
20. _____

Inch-Based Micrometer

The reading is composed of:

4 Large graduations or 4 x 0.100   = 0.400
2 Small graduations or 2 x 0.025   = 0.050
8 Graduations on the thimble or 8 x 0.001 = 0.008
$$\overline{ 0.458''.}$$

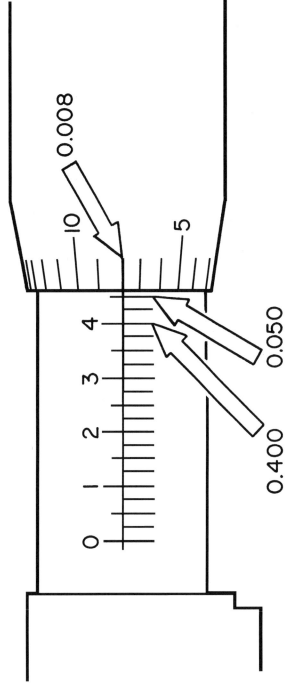

Inch-Based Vernier Micrometer

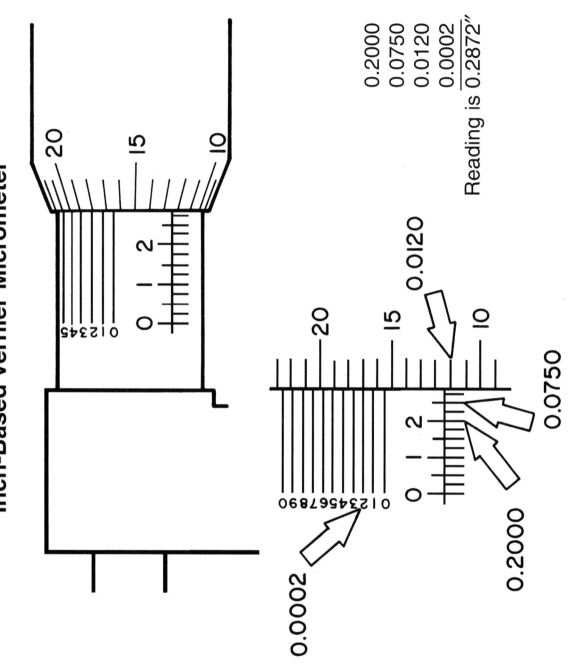

0.2000
0.0750
0.0120
0.0002
Reading is 0.2872"

Metric Micrometer

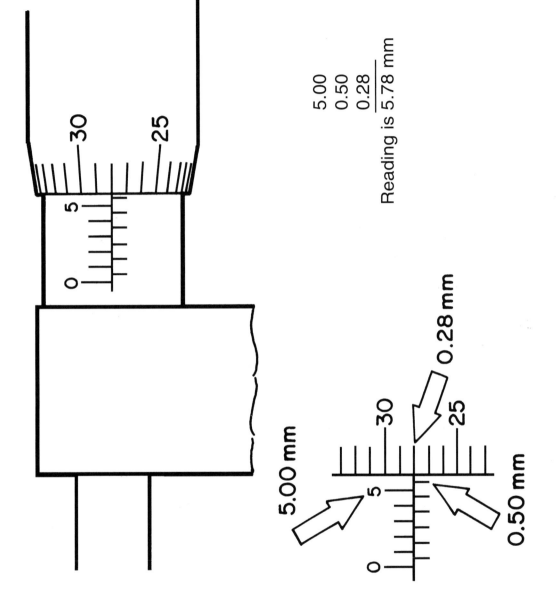

5.00
0.50
0.28
────
Reading is 5.78 mm

5.00 mm

0.28 mm

0.50 mm

Metric-Based Vernier Micrometer

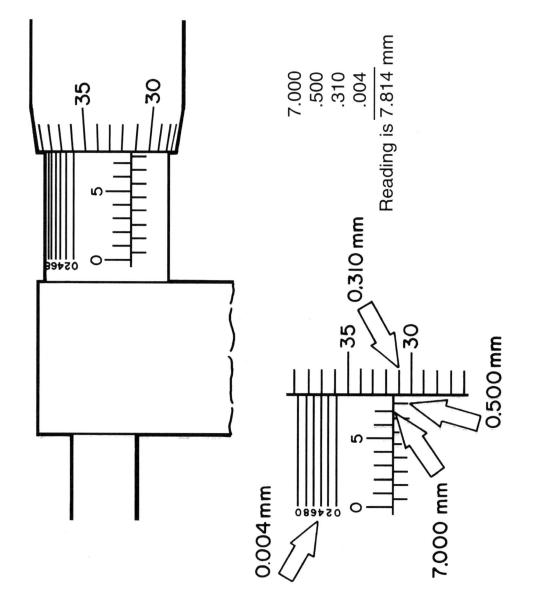

7.000
.500
.310
.004
Reading is 7.814 mm

0.310 mm

0.004 mm

0.500 mm

7.000 mm

25-Division Inch-Based Vernier Caliper

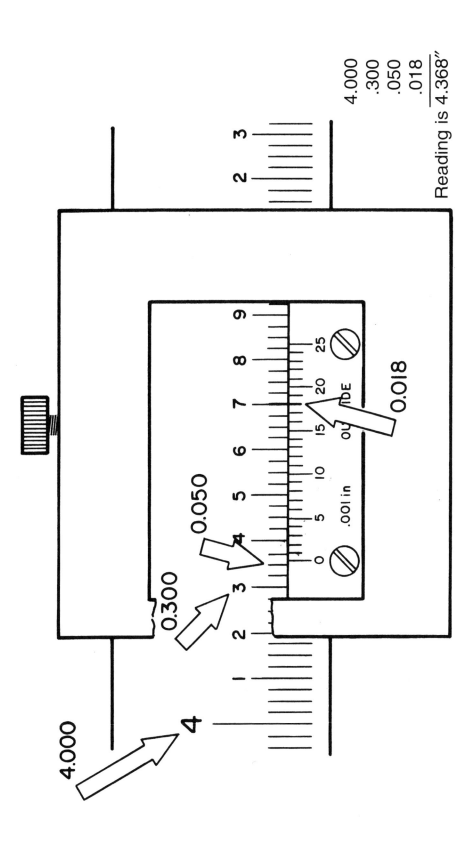

4.000
.300
.050
.018
Reading is 4.368"

25-Division Metric Vernier Caliper

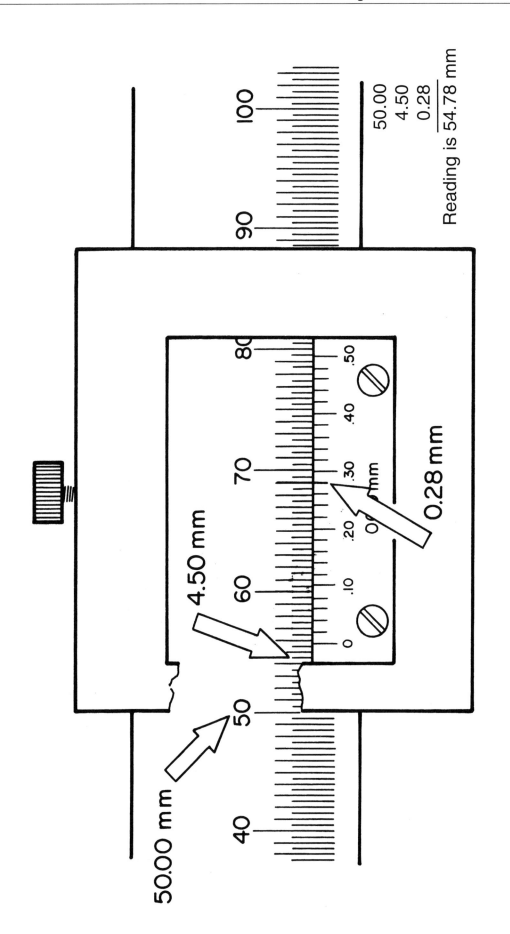

50.00 mm

4.50 mm

0.28 mm

50.00
4.50
0.28
Reading is 54.78 mm

4-9

50-Division Inch-Based Vernier Caliper

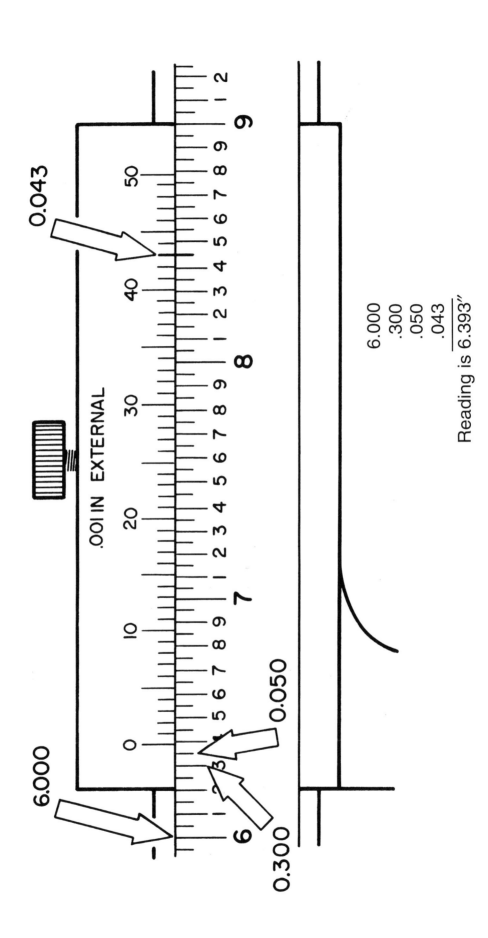

6.000
.300
.050
.043
Reading is 6.393"

50-Division Metric Vernier Caliper

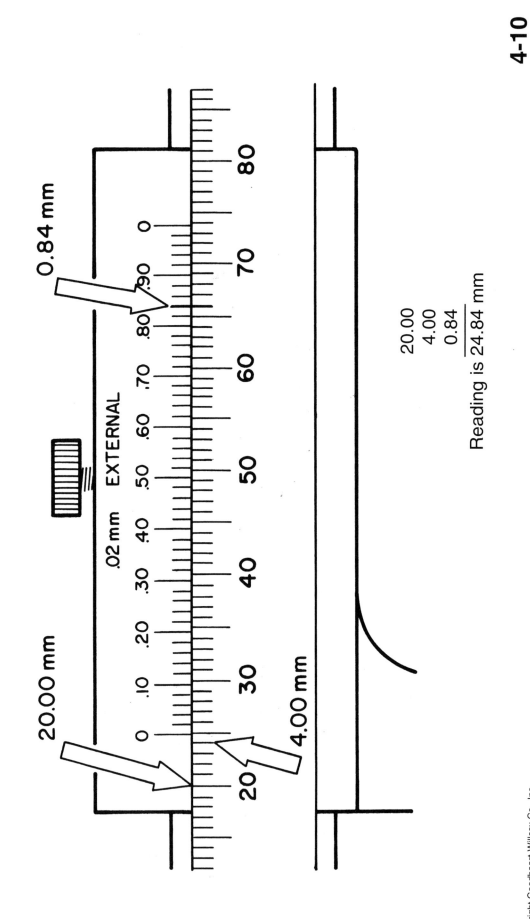

20.00
4.00
0.84
Reading is 24.84 mm

Universal Bevel Protractor

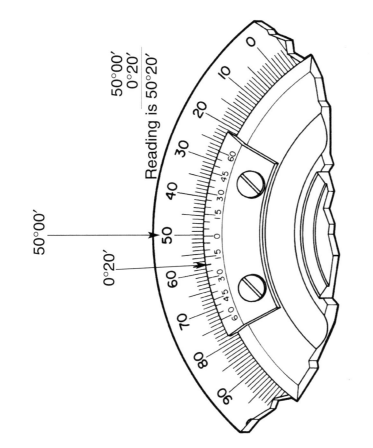

50°00'
0°20'

50°00'
0°20'
Reading is 50°20'

To read the protractor, note the number of degrees that can be read up to the "0" on the Vernier plate. To this, add the number of minutes indicated by the line beyond the "0" on the Vernier plate that aligns exactly with a line on the dial.

The "0" is past the 50° mark, and the Vernier scale aligns at the 20' mark. Therefore, the measurement is 50°20'.

4-11

Measurement

Name: _____ Date: _____ Score: _____

1. Make readings from the rules shown below.

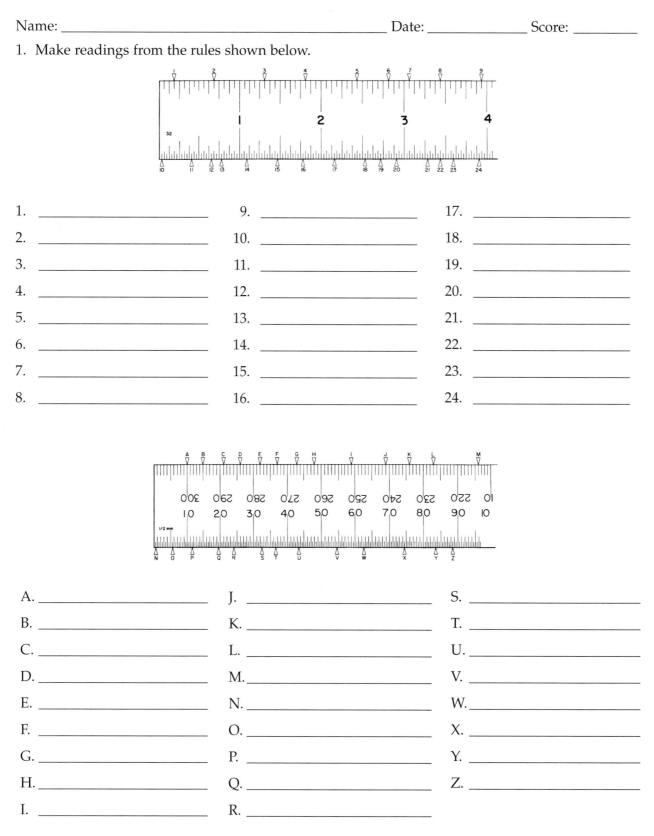

1. _____
2. _____
3. _____
4. _____
5. _____
6. _____
7. _____
8. _____

9. _____
10. _____
11. _____
12. _____
13. _____
14. _____
15. _____
16. _____

17. _____
18. _____
19. _____
20. _____
21. _____
22. _____
23. _____
24. _____

A. _____
B. _____
C. _____
D. _____
E. _____
F. _____
G. _____
H. _____
I. _____

J. _____
K. _____
L. _____
M. _____
N. _____
O. _____
P. _____
Q. _____
R. _____

S. _____
T. _____
U. _____
V. _____
W. _____
X. _____
Y. _____
Z. _____

4-12

(continued)

Name: _____

2. Make readings from the Vernier scales shown below.

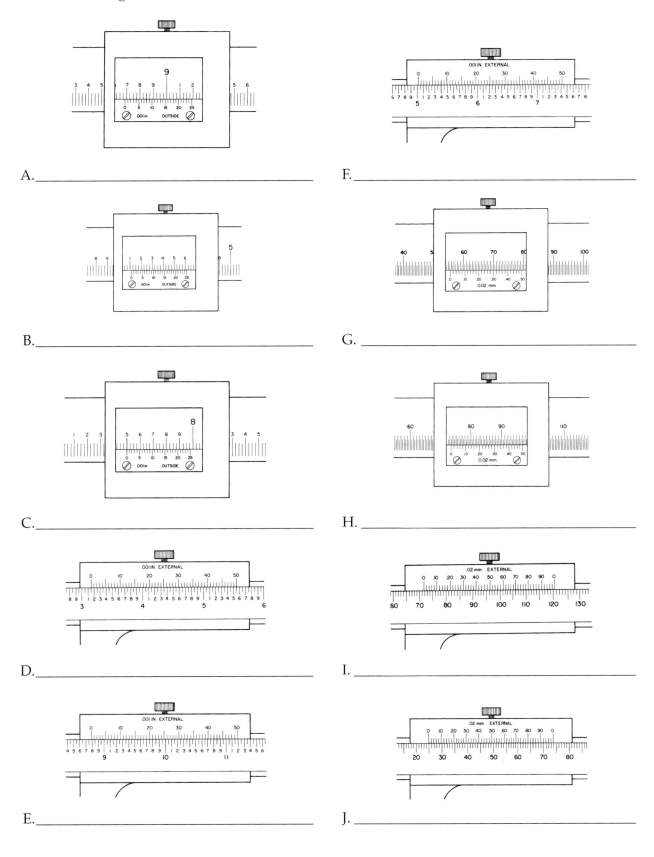

A._____

B._____

C._____

D._____

E._____

F._____

G._____

H._____

I._____

J._____

4-12

(continued)

Name: _____

• Answer the following questions as they pertain to measurement.

3. The micrometer is nicknamed _____.

    3. _____

4. One-millionth part of a standard inch is known as a _____.

    4. _____

5. One-millionth part of a meter is known as a _____.

    5. _____

6. A micrometer is capable of measuring accurately to the _____ and _____ part of standard inch and (in metric versions) to _____ and _____ millimeters.

    6. _____

       _____

       _____

       _____

7. The Vernier caliper has several advantages over the micrometer. List two of them.

_____

_____

8. A Vernier caliper can measure to the _____ part of the inch and (in the metric version) to _____ millimeters.

    8. _____

       _____

9. List six precautions that must be observed when using a micrometer or Vernier caliper.

_____

_____

_____

_____

_____

_____

10. The Vernier-type tool for measuring angles is called a _____.

    10. _____

11. How does a double-end cylindrical plug gage differ from a step plug gage? _____

_____

_____

_____

12. A ring gage is used to check whether _____ are within the specified _____ range.

    12. _____

       _____

13. Gage blocks are often referred to as _____ blocks.

    13. _____

Name: _____

14. An air gage employs air pressure to measure deep internal   14. _____
openings and hard-to-reach shaft diameters. It operates on
the principle of:
a. Air pressure leakage between the plug and hole walls.
b. The amount of air pressure needed to insert the tool properly in the hole.
c. Amount of air pressure needed to eject the gage from the hole.
d. All of the above.
e. None of the above.

15. The dial indicator is available in two basic types. List them.

_____

_____

16. What are some uses for the dial indicator? _____

_____

_____

_____

17. Name the measuring device that employs light waves as a measuring standard.

_____

18. The _____ is used for production inspection. An   18. _____
enlarged image of the part is projected on a screen where
it is superimposed upon an accurate drawing.

19. The pitch of a thread can be determined with a _____.   19. _____

20. Of what use are fillet and radius gages? _____

_____

_____

21. What are helper measuring tools? _____

_____

_____

22. How is a telescoping gage used? _____

_____

_____

_____

_____

4-12
(continued)

Name: _____

23. Make readings from the micrometer illustrations.

A. _____   G. _____

B. _____   H. _____

C. _____   I. _____

D. _____   J. _____

E. _____   K. _____

F. _____   L. _____

A

B

C

D

E

F

G

H

I

J

K

L

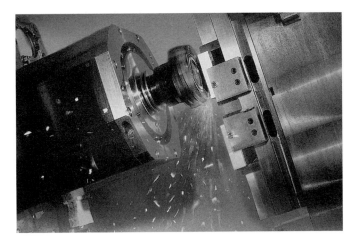

Chapter 5

Layout Work

LEARNING OBJECTIVES

After studying this chapter, students will be able to:
- ○ Explain why layouts are needed.
- ○ Identify common layout tools.
- ○ Use layout tools safely.
- ○ Make basic layouts.
- ○ List safety rules for layout work.

INSTRUCTIONAL MATERIALS

Text: pages 81–90
    Test Your Knowledge Questions, page 89
Workbook: pages 33–38
Instructor's Resource: pages 93–100
Guide for Lesson Planning
    Research and Development Ideas
    Reproducible Masters:
        5-1 Typical Layout Problem
        5-2 Steps in Making the Layout
        5-3 Test Your Knowledge Questions
    Color Transparency (Binder/CD only)

GUIDE FOR LESSON PLANNING

The chapter serves as an introduction to basic layout tools and materials as well as making a layout. Prepare for the lesson by having the following equipment available:

- Sections of clean metal to demonstrate layout techniques.
- Layout dye, scribers, hermaphrodite caliper, divider, surface gage, selection of squares, combination set, hammer, and punches.

For a demonstration on precision layout work, have the following equipment available:

- Vernier height gage, right angle plate, parallels, V-blocks, straight edge, Vernier bevel protractor, and surface plate.

Have students read and study the chapter paying attention to the illustrations. Discuss and demonstrate the layout tools they will be using. This should include the following:

- Why layouts are necessary.
- Safe use of layout tools.
- How to prepare metal for layout.
- Proper use of various layout tools.
- Steps in making a simple layout.
- Laying out angles.
- The use of parallels, V-blocks, and angle plate in layout work.
- Proper way to use and care for Vernier type layout tools.
- Care of the surface plate.
- Safety rules to be observed when making layouts.

Technical Terms

Review the terms introduced in the chapter. New terms can be assigned as a quiz, homework, or extra credit. The terms are also listed at the beginning of the chapter.

divider
hardened steel square
layout dye
plain protractor
reference line
scriber
straightedge
surface gage
surface plates
V-blocks

Review Questions

Assign *Test Your Knowledge* questions. Copy and distribute Reproducible Master 5-3 or have students use the questions on page 89 and write their answers on a separate sheet of paper.

Workbook Assignment

Assign Chapter 5 of the *Machining Fundamentals Workbook*.

Research and Development

Discuss the following topics in class or have students complete projects on their own.

1. Make a display panel showing samples of the various layout fluids used by industry. Use a clear plastic spray to prevent the scribed lines from rusting and the coatings from rubbing off.

2. Prepare a sample of a good layout job. Develop it into a bulletin board display. Use colored twine or yarn running from the sample to printed notations explaining the various aspects that indicate a good layout job.

3. Write a paper on how surface plates are made. Secure literature from the various manufacturers to illustrate the paper. Also include:

   • How surface plate grades are determined.

   • Why cast iron, steel, and granite are used to make them rather than other materials.

   • How to take care of the surface (maintain accuracy, keep it clean, etc.)

4. Prepare a series of overhead projector transparencies, 35 mm slides, or a video to show the correct sequence for producing a good layout job.

TEST YOUR KNOWLEDGE ANSWERS, Page 89

1. Layout dye.
2. To locate and mark out lines, circles, arcs, and points for drilling holes. They show machinist where to machine.
3. scriber
4. divider
5. trammel
6. Lines will rub off and would be too wide.
7. surface plate
8. Evaluate individually. Refer to Section 5.2.
9. V-blocks
10. straightedge
11. center head
12. Vernier protractor
13. prick, center
14. Any three of the following:

    Never carry an open scriber, divider, trammel, or hermaphrodite caliper in your pocket.

    Always cover sharp points with a cork when the tool is not being used.

    Wear goggles when grinding scriber points.

    Get help when you must move heavy items, such as angle plates or V-blocks.

    Remove all burrs and sharp edges from stock before starting layout work.

WORKBOOK ANSWERS, Pages 33–38

1. layout dye
2. d. All of the above.
3. b. can be used to locate the center of irregularly shaped stock
4. e. None of the above.
5. a. trammel
6. b. parallel
7. flatness
8. c. plain protractor
9. a. universal bevel
10. b. protractor depth gage
11. a. double square
12. d. All of the above.
13. bevel protractor
14. Any order: never carry an open scriber, divider, trammel, or hermaphrodite caliper

in your pocket; always cover sharp points with a cork when the tool is not being used; wear goggles when grinding scriber points; get help when you must move heavy items, such as angle plates or V-blocks; remove all burrs and sharp edges from stock before starting layout work.

15. Trammel
16. Protractor depth gage
17. Universal bevel
18. Evaluate individually.
19. Rule, scribe, square, divider, prick punch, center punch, hammer.
20. Evaluate individually. Refer to Section 5.4.

Typical Layout Problem

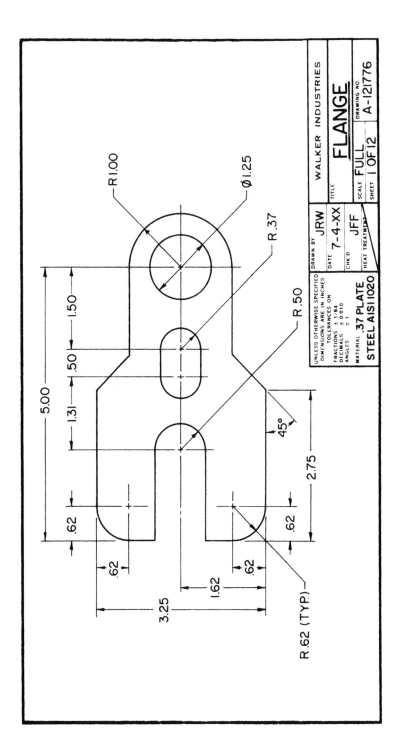

5-2

Steps in Making the Layout

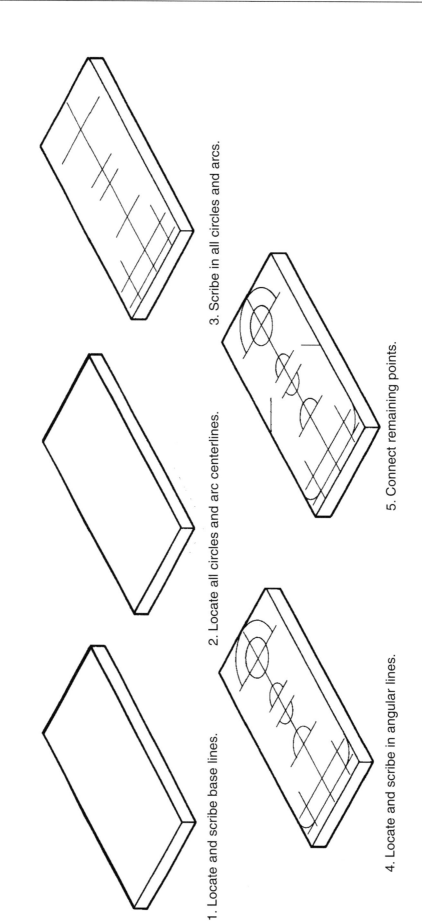

1. Locate and scribe base lines.

2. Locate all circles and arc centerlines.

3. Scribe in all circles and arcs.

4. Locate and scribe in angular lines.

5. Connect remaining points.

Layout Work

Name: _____ Date: _____ Score: _____

1. What is used to make layout lines easier to see? 1. _____

2. Why are layout lines used?_____

_____

_____

3. Straight layout lines are drawn with a _____. 3. _____

4. Circles and arcs are drawn on work with a _____. 4. _____

5. Large circles and arcs are drawn with a _____. 5. _____

6. What is wrong with using a pencil to make layout lines 6. _____
on metal?

_____

7. A _____ _____ is the flat granite or steel surface used for 7. _____
layout and inspection work.

8. What layout operations can be performed with a combination set?

_____

_____

_____

_____

9. Round stock is usually supported on _____ for layout 9. _____
and inspection.

10. Long flat surfaces can be checked for trueness with a 10. _____
_____.

11. The center of round stock can be found quickly with the 11. _____
_____ and rule of a combination set.

12. Angular lines that must be very accurate should be laid 12. _____
out with a _____.

13. The _____ punch has a sharper point than the _____ 13. _____
punch.

14. List three safety precautions that you should observe when doing layout work.

_____

_____

_____

Chapter **6**

Hand Tools

LEARNING OBJECTIVES

After studying this chapter, students will be able to:
- ⭕ Identify the most commonly used machine shop hand tools.
- ⭕ Select the proper hand tool for the job.
- ⭕ Maintain hand tools properly.
- ⭕ Explain how to use hand tools safely.

INSTRUCTIONAL MATERIALS

Text: pages 91–126
    Test Your Knowledge Questions,
    pages 123–125
Workbook: pages 33–38
Instructor's Resource: pages101–120
    Guide for Lesson Planning
    Research and Development Ideas
    Reproducible Masters:
        6-1 Torque Measurement
        6-2 Mounting Work for Hand Sawing
        6-3 Starting a Hand Reamer
        6-4 Using a Hand Reamer
        6-5 Specifications for Thread Sizes
        6-6 Inch-Based and Metric-Based Threads
        6-7 Thread Nomenclature
        6-8 Starting a Die
        6-9 Cutting Threads to a Shoulder
        6-10 Test Your Knowledge Questions
    Color Transparencies (Binder/CD only)

GUIDE FOR LESSON PLANNING

    This chapter serves to present the numerous hand tools that a machinist will use as well as their proper handling and care. Since this chapter is extensive, it is recommended that it be divided into several parts. It can be taught as a series of short lessons in which students can become actively involved.

Student Presentations

    Assign pairs of students or request volunteers to demonstrate and explain the proper care and use for the family of tools on which they are providing instructions. The following preparations should be made before each presentation:
- All tools in safe working condition.
- Additional tools in reserve.
- Demonstration clearly visible to all.
- All safety precautions taken.
- Have students read and study the assignment.

    Students should furnish a written outline for the lesson they are to present. Aid them in preparing their topic for discussion and questions. In addition to becoming more involved in the class, students will gain experience in preparing and giving presentations.

    An outline for hand threading, for example, would include the following:

I. Objectives

A. After studying this topic, students should be able to:
1. Understand how threads are specified on drawings.
2. Explain thread nomenclature.
3. Select the proper tap(s) and tap wrench for each job.
4. Determine the correct tap drill size for specified thread to be tapped.
5. Adjust a die for different classes of fits.
6. Use, clean, and store threading tools properly.
7. Observe hand threading safety.

II. Instructional Aids

A. Text pages 114–122
B. Reproducible Masters:
   3-3 How Threads are Depicted on Drawings
   6-5 Specifications for Thread Sizes
   6-6 Inch-Based and Metric-Based Threads
   6-7 Thread Nomenclature
   6-8 Starting a Die
   6-9 Cutting Threads to a Shoulder

A list of necessary equipment should also be prepared. For example, the list below includes equipment necessary to demonstrate hand threading.

- Examples of UNC, UNF, and metric threaded sections (bolts, nuts, threaded rods, etc.)
- Different tap sets
- Different size tap wrenches
- Material drilled for tapping
- A selection of dies
- Different size die holders
- Stock for threading with a die
- Cutting fluid

The demonstration area should be clearly visible to all students.

Have students read and study the material prior to the presentation, paying special attention to the illustrations. Allow students to use the Reproducible Masters as overhead transparencies or handouts, as they discuss and demonstrate their topic. The following list includes some areas that should be covered.

- How threads are depicted on drawings
- How UNC and UNF threads of the same size differ
- Thread nomenclature
- Why there are taper, plug, and bottom taps
- Tap drills and their importance
- Proper way to tap a hole
- Why cutting fluids are necessary
- Removing broken taps
- The proper way to cut threads with a die
- Advantages of using adjustable dies
- Precautions to be taken when hand threading
- How to clean and store hand threading tools
- Importance of washing hands thoroughly after hand threading

A review of the demonstrations will provide students the opportunity to ask questions.

Technical Terms

Review the terms introduced in the chapter. New terms can be assigned as a quiz, homework, or extra credit. These terms are also listed at the beginning of the chapter.
   abrasive
   American National Thread System
   blind hole
   classes of fits
   foot-pounds
   newton meters
   number sizes
   safe edges
   torque
   Unified System

Review Questions

Assign *Test Your Knowledge* questions. Copy and distribute Reproducible Master 6-10 or have students use the questions on pages 123–125 and write their answers on a separate sheet of paper.

Workbook Assignment

Assign Chapter 6 of the *Machining Fundamentals Workbook*.

Research and Development

Discuss the following topics in class or have students complete projects on their own.

1. Industry makes considerable use of the

pneumatic chisel. Secure information on this tool for a bulletin board display and, if possible, borrow the tool for examination from a local industry.

2. Design a safety poster that shows the correct way to use a chisel.

3. Secure samples of the various types of hacksaw blades used for hand sawing. Prepare them as a bulletin board display.

4. Design and produce a series of safety posters on the file that illustrate the following unsafe practices:
   - Using a file as a pry.
   - File used without a handle.
   - File used as a hammer.

5. Design a panel that shows the file in various stages of manufacturer. Secure samples.

6. Inspect the files in your shop. Clean and repair or replace damaged file handles. Make a new file rack if the present rack is badly worn.

7. Examine the screwdrivers in your shop. Repair or regrind the tools as needed.

8. There are many other types of wrenches not covered in this unit. Prepare a paper featuring these wrenches. Include drawings. Reproduce the report for distribution to the class.

9. Give a demonstration on the proper way to use a torque-limiting wrench.

10. Contact various tool manufacturers for information on how wrenches are manufactured. Prepare a bulletin board display with the material.

11. Repair and lubricate all adjustable wrenches in the shop.

12. Make a safety poster illustrating the proper way to use a wrench.

13. Prepare a sample block of metal that can be used to show the difference between a drilled hole and a reamed hole.

14. Develop and construct displays that show:
   - Samples of various abrasive materials.
   - A flow chart showing how synthetic abrasives are manufactured. If possible, secure samples of the raw materials.
   - Metal samples in various stages of polishing. Spray them with lacquer or acrylic plastic to prevent rust.

15. Set up an experiment to determine what abrasive materials are best for aluminum, brass, cast iron, and tool steel. The experiment should include the quantity of material removed within a specified period of time; surface finish of the completed piece; degree of clogging, if any, of the abrasive cloth; and the effect lubricating oil has on the surface finish. Abrasives of similar grade value must be used if tests are to be valid.

16. Give a demonstration on the different methods for removing broken taps. Industry often uses a technique that erodes the tap electrically, permitting the parts to be removed easily. Secure information on this process for a bulletin board display.

17. Prepare a study on the accuracy of hand reamers. Make sample holes and measure them to determine whether they are within acceptable limits. Does the application of cutting fluid affect the size of a reamed hole?

18. Demonstrate the proper way to tap a blind hole.

19. Demonstrate the correct way to run a thread down to a shoulder.

TEST YOUR KNOWLEDGE ANSWERS, Pages 123–125

1. Solid base and swivel base.

2. By the width of the jaws.

3. soft metal caps (copper, brass or aluminum)

4. Avoid letting the vise handle or work project into aisle beside bench.

5. C-clamp, parallel clamp

6. They can be opened wider at the hinge pin to grip larger size work.

7. Permits them to cut flush with the work surface.

8. Student answers will vary but may include any three of the following: *never* using as a substitute for a wrench; not trying to cut metal sizes that are too large, or work that has been heat-treated; not applying additional leverage to the handles; cleaning and oiling them; storing in a clean, dry place; not throwing them in a drawer or tool box with other tools; using pliers that are large enough for the job.

9. Pliers that can be adjusted to various size work and can clamp tightly on the work. They have a quick release.

10. They permit tightening a threaded fastener or part for maximum holding power without danger of fastener or part failing, or causing work to warp or spring out of shape.

11. Work equally well in either direction but for safety, they should be pulled.

12. Pipe wrench, monkey wrench, and regular adjustable wrench.

13. Any three of the following: the movable jaw should face the direction the fastener is to be rotated; the thumbscrew should be adjusted so the jaws fit the bolt head or nut snugly; an extension should not be used for additional leverage; should never hammer on the handle to loosen a stubborn fastener; the smallest wrench that will fit the fastener should be used.

14. pipe, damage

15. Socket wrenches are box-like and are made with a tool head-socket (opening) that fits many types of handles (either solid bar or ratchet type).

16. Pin, hook, and end spanner wrenches.

17. larger

18. Any five of the following: always pull on a wrench; never push; select a wrench that fits properly; never hammer on a wrench to loosen a stubborn fastener; rather than lengthening a wrench handle for additional leverage, use a larger wrench; clean any grease or oil off the handle and the floor in the work area before using a wrench; and never try to use a wrench on moving machinery.

19. Standard has a wedge-shaped tip. Phillips has an X-shaped tip.

20. e. Tip is similar to that of a Phillips head screwdriver.

21. a. Has a flattened wedge-shaped tip.

22. f. Has an insulated handle.

23. c. Has a square shank to permit additional force to be applied with a wrench.

24. g. Is short and is used when space is limited.

25. b. Moves the fastener on the power stroke, but not on the return stroke.

26. Student answers will vary but may include any three of the following: screwdrivers should not be used as a substitute for a chisel, hammered on, or used as a pry bar; should always wear safety goggles when grinding screwdriver tips; burred heads (on screws) should be replaced or the burrs

removed with a file or abrasive cloth; the screwdriver should have an insulated handle and the power should always be turned off before working on electrical equipment; screwdrivers should not be carried in pockets.

27. By the weight of the head.

28. Used to strike heavy blows where steel faced hammers would damage or mar the work.

29. Student answers will vary but may include any three of the following: never strike two hammers together; do not use a hammer unless the head is on tightly and the handle is in good condition; do not "choke up" too far on the handle when striking a blow, or you may injure your knuckles; strike each blow squarely, or the hammer may glance off of the work and injure you or someone working nearby; place a hammer on the bench carefully so that it doesn't fall and cause a painful foot injury, or damage precision tools on the bench.

30. mushroomed, grinding

31. removing rivets

32. Flat, cape, round nose, and diamond point chisels.

33. different length blades

34. 40, 50

35. Vibration and chatter are eliminated. They cause blade to dull rapidly.

36. As a blade dulls the slot made by it becomes narrower. To continue in this slot will cause a new blade to bind and be ruined in a few strokes.

37. three or more, teeth will straddle work and be broken off

38. Mount it between two sections of wood.

39. Place soft wood blocks between the vise jaws and the work to prevent marring the exterior surface of the tubing and insert a snug-fitting wooden dowel into the tubing.

40. file card/brush, your hand

41. Single-cut, double-cut, rasp, and curved tooth.

42. Flat, pillar, square, 3-square, knife, half-round, crossing, round.

43. Student answers will vary but may include any three of the following: never use a file without a handle; files should be cleaned with a file card, not your hand; files should not be cleaned by slapping it on the bench, since it may shatter; files are very brittle and

should never be used for prying tasks; a piece of cloth should be used to clean the surface being filed, not your bare hand; you should never hammer on or with a file because it could shatter.

44. When a hole must be finished accurately to size and with a very smooth surface finish.

45. 0.005″ to 0.010″ (0.15 mm to 0.25 mm)

46. tap, die

47. e. None of the above.

48. tap drill

49. UNC has fewer threads per inch for a given diameter than the UNF of the same diameter.

50. In order: taper, plug, and bottom.

51. Neither. It must be the same size as the desired threads.

52. tap wrench, die stock

53. An abrasive is any hard, sharp material that is used to cut or grind away another material.

54. Machined

WORKBOOK ANSWERS, Pages 39–46

1. d. All of the above.

2. a. using the handle to turn the heavy screw

3. paper

4. b. overall length of the tool

5. e. Both b and c.

6. Under no condition should the handle be lengthened for additional leverage.

7. Pulled. Pushing any wrench is considered dangerous.

8. a. the movable jaw should face the direction the fastener is to be rotated

9. c. box wrench

10. Torque is the amount of turning or twisting force applied to a threaded fastener or part.

11. d. All of the above.

12. Hook, pin, and end.

13. Any four of the following: always pull on a wrench; never push; select a wrench that fits properly; never hammer on a wrench to loosen a stubborn fastener; rather than lengthening a wrench handle for additional leverage, use a larger wrench; clean any grease or oil off the handle and the floor in the work area before using a wrench; never try to use a wrench on moving machinery.

14. A. Standard
    B. Phillips
    C. Clutch
    D. Square
    E. Torx
    F. Hex

15. A. For general cutting.
    B. To cut grooves.
    C. To cut radii and round grooves.
    D. For squaring corners.

16. entire length

17. b. chatter and vibration will dull the teeth

18. a. cut will be too narrow for the new blade

19. a. the teeth will straddle the section being cut and snap off

20. c. general outline and cross section

21. A. Single cut
    B. Double cut
    C. Rasp
    D. Curved tooth

22. draw

23. surface finish

24. c. expansion hand

25. b. spiral-fluted

26. Student answers will vary but may include any three of the following: to prevent injury, remove all burrs from holes; never use your hands to remove chips and cutting fluid from the reamer (use a piece of cotton waste); store reamers carefully so they do not touch one another, they should never be stored loose or thrown into a drawer with other tools; clamp work solidly before starting to ream; do not use compressed air to remove chips and cutting fluid or to clean a reamed hole.

27. e. None of the above.

28. In order: taper, plug, and bottom.

29. Diameter must be same size as threads.

30. Any of the following: too little or lack of cutting oil, dull die cutters; stock too large for threads being cut; die not started square; one set of cutters could be upside down when using a two part die.

31. b. used to wear away another material

32. Emery

33. Silicon carbide

34. A–F, evaluate individually.

35. Evaluate individually.

Torque Measurement

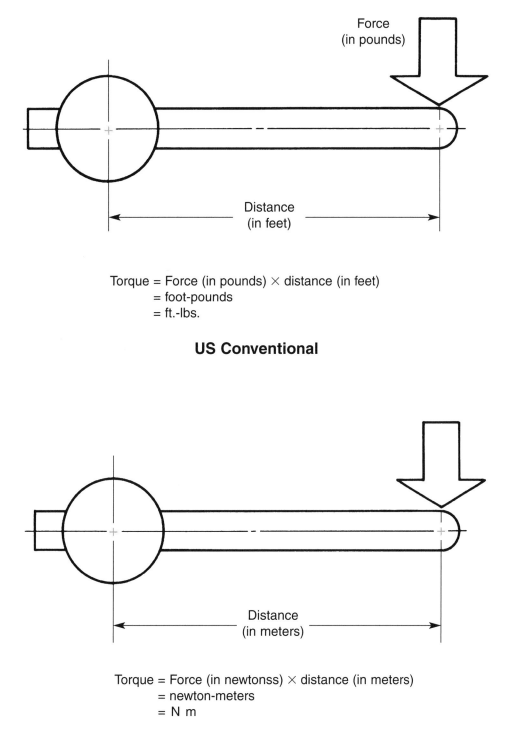

Torque = Force (in pounds) × distance (in feet)
= foot-pounds
= ft.-lbs.

US Conventional

Torque = Force (in newtonss) × distance (in meters)
= newton-meters
= N m

SI Metric

6-1

Mounting Work for Hand Sawing

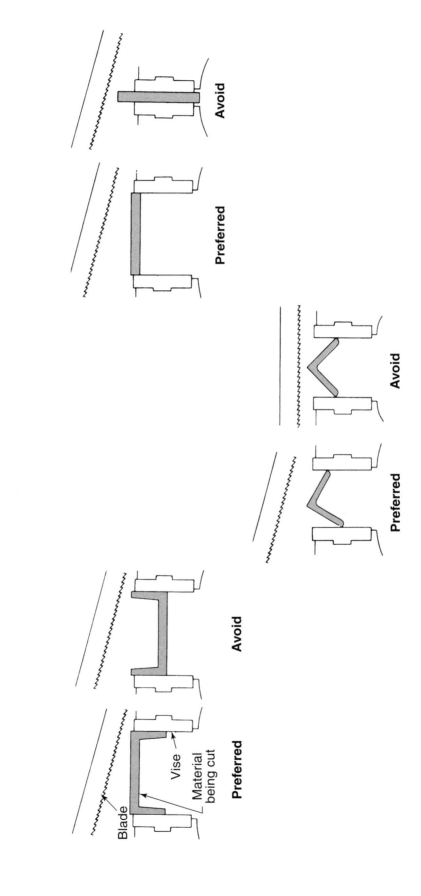

Avoid

Preferred

Avoid

Preferred

Avoid

Preferred

Blade

Vise

Material being cut

Starting a Hand Reamer

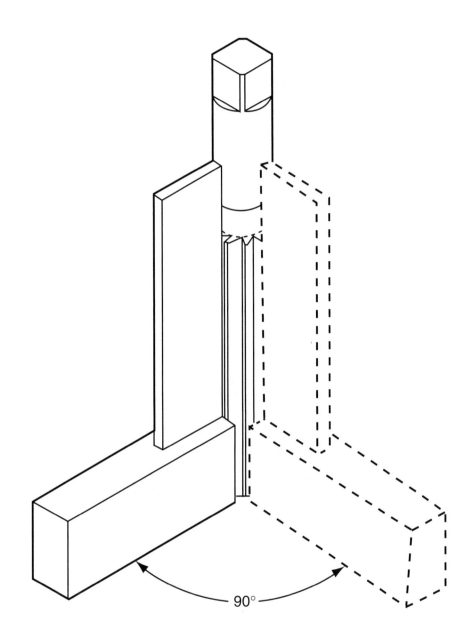

90°

Always make sure that the reamer is square with the work.

6-3

Using a Hand Reamer

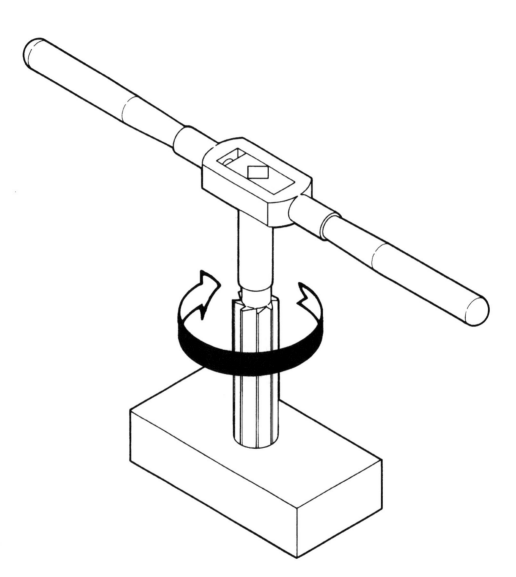

Always turn a hand reamer in a clockwise direction.

Specifications for Thread Sizes

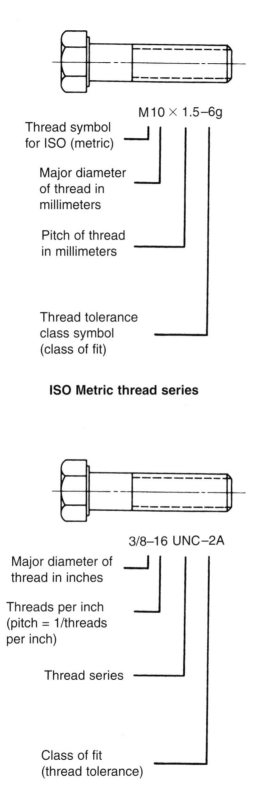

M10 × 1.5–6g

Thread symbol
for ISO (metric)

Major diameter
of thread in
millimeters

Pitch of thread
in millimeters

Thread tolerance
class symbol
(class of fit)

ISO Metric thread series

3/8–16 UNC–2A

Major diameter of
thread in inches

Threads per inch
(pitch = 1/threads
per inch)

Thread series

Class of fit
(thread tolerance)

Unified National coarse thread series

Inch-Based and Metric-Based Threads

ISO Metric Thread Series   Unified National Coarse Thread Series

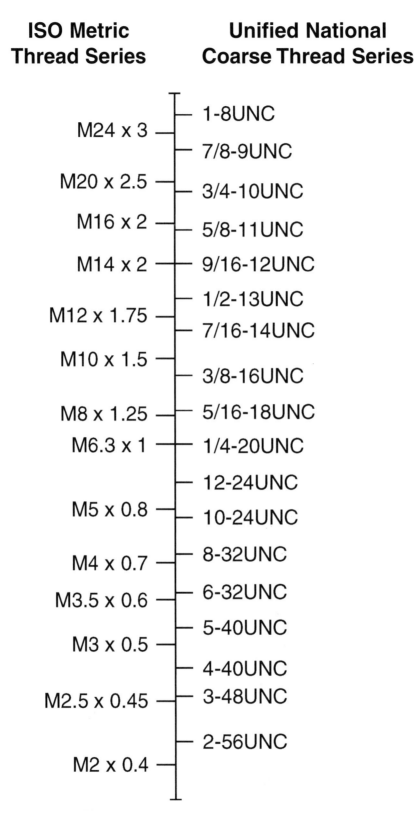

ISO Metric Thread Series	Unified National Coarse Thread Series
	1-8UNC
M24 x 3	7/8-9UNC
M20 x 2.5	3/4-10UNC
M16 x 2	5/8-11UNC
M14 x 2	9/16-12UNC
	1/2-13UNC
M12 x 1.75	7/16-14UNC
M10 x 1.5	3/8-16UNC
M8 x 1.25	5/16-18UNC
M6.3 x 1	1/4-20UNC
	12-24UNC
M5 x 0.8	10-24UNC
M4 x 0.7	8-32UNC
M3.5 x 0.6	6-32UNC
	5-40UNC
M3 x 0.5	4-40UNC
M2.5 x 0.45	3-48UNC
	2-56UNC
M2 x 0.4	

A comparison of ISO metric coarse and Unified Coarse (UNC) inch-based thread sizes. Even though several of them seem to be the same size, they are not interchangeable (one cannot be substituted for the other).

Thread Nomenclature

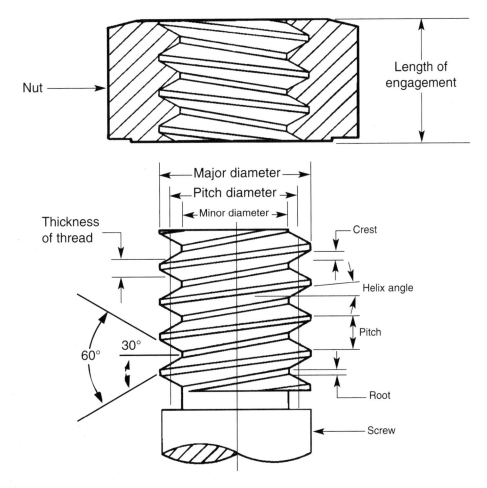

Nut

Length of engagement

Major diameter

Pitch diameter

Minor diameter

Thickness of thread

Crest

Helix angle

60°

30°

Pitch

Root

Screw

Starting a Die

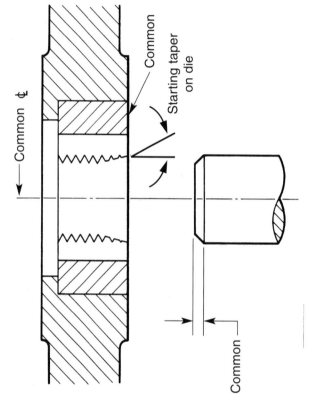

Common ℄

Common

Starting taper
on die

Common

A die will start more easily if a small chamfer is cut or ground on the end of the shaft to be threaded. Section through die and die stock shows proper way to start threads.

Cutting Threads to a Shoulder

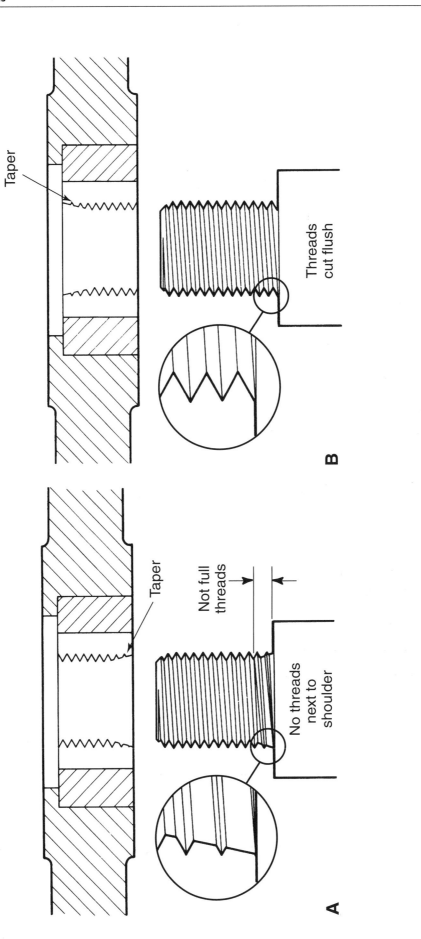

After die has been run down as far as possible, the die is reversed. When rotated down the shaft, it will cut threads almost flush with shoulder. A—Running die down normally. B—Reversing die to cut flush.

6-9

Hand Tools

Name: _____ Date: _____ Score: _____

1. List two variations of the machinist's vise. _____

   _____

2. How is vise size determined? _____

   _____

3. Work held in a vise can be protected from damage by the   3. _____
   jaw serrations if _____ are placed over the jaws.

4. To prevent injuries, what should be avoided when mounting work in a vise? _____

   _____

5. Work is often held together with a _____ and/or _____   5. _____
   while being machined or worked on.
                                                                  _____

6. How do combination pliers have an advantage over many other types of pliers? _____

   _____

7. Why are the cutting edges on diagonal pliers set at an angle? _____

   _____

8. List three ways of extending the working life of pliers.

   _____

   _____

   _____

9. What are adjustable clamping pliers? _____

   _____

   _____

10. Of what use are torque-limiting wrenches? _____

    _____

    _____

11. Do torque-limiting wrenches give a more accurate reading when they are pushed or when they
    are pulled? _____

    _____

    _____

12. Several different wrenches can be classified as adjustable wrenches. Name three. _____

    _____

6-10

(continued)

Name: _____

13. List three points that should be observed when using an adjustable wrench. _____

_____

_____

_____

14. Round work can be gripped with a _____ wrench. Its   14. _____
main disadvantage is that the jaws will probably _____
the work.   _____

15. Describe socket wrenches. _____

_____

_____

16. What wrenches are employed to turn flush and recessed types of threaded fasteners? The fasteners
have slots or holes to receive the wrench lugs. _____

_____

17. Rather than lengthen the wrench handle for additional   17. _____
leverage, it is better to use a _____ wrench.

18. List five safety precautions that should be observed when using a wrench.

_____

_____

_____

_____

_____

19. What is the difference between a standard screwdriver tip and a Phillips screwdriver tip?

_____

• Match each phrase in the right column with the correct screwdriver name in the left column.

_____ 20. Pozidriv®	a.	Has a flattened wedge-shaped tip.
_____ 21. Standard	b.	Moves the fastener on the power stroke, but not on the return stroke.
_____ 22. Electrician	c.	Has a square shank to permit additional force to be applied with a wrench.
_____ 23. Heavy-duty	d.	Useful when handling small screws.
_____ 24. Stubby	e.	Tip is similar to that of a Phillips head screwdriver.
_____ 25. Ratchet	f.	Has an insulated handle.
	g.	Is short and is used when space is limited.

6-10
(continued)

Name: _____

26. List three safety precautions that should be observed when using a screwdriver.

_____

_____

_____

27. How is the size of a ball-peen hammer determined?

_____

28. Why are soft-face hammers and mallets used in place of a ball-peen hammer? _____

_____

29. List three safety precautions that should be observed when using striking tools.

_____

_____

_____

30. There are few things more dangerous than a chisel with a head that has become _____ from use. This danger can be removed by _____.

30. _____

_____

31. The chisel is an ideal tool for _____ _____.

31. _____

32. List the four general types of cold chisels. _____

_____

33. The standard hacksaw is designed to accommodate _____.

33. _____

34. A hacksaw cuts best at about _____ to _____ strokes per minute.

34. _____

_____

35. Why should work be mounted solidly and close to the vise before cutting with a hacksaw?

_____

_____

36. Should a blade break or dull before completing a cut, you should *not* continue in the same cut with a new blade. Why? _____

_____

_____

37. The number of teeth per inch on a hacksaw blade has an important bearing on the shape and kind of metal being cut. At least _____ or _____ should be cutting at all times, otherwise _____.

_____

_____

Name: _____

38. What is the best way to hold thin metal for hacksawing?_____

_____

39. What is the best way to hold thin wall tubing for hacksawing?

_____

_____

40. Files are cleaned with a _____, never with _____.   40. _____

_____

41. Files are classified according to the cut of their teeth. List the four cuts. _____

_____

42. What are the most commonly used file shapes? _____

_____

43. List three safety precautions that should be observed when files are used. _____

_____

_____

_____

44. When is reaming done?_____

_____

45. How much stock should be left in a hole for hand   45. _____
    reaming?

46. A _____ is used to cut internal threads. External threads   46. _____
    are cut with a _____.
    _____

47. The hole to be tapped must be: 47. _____
    a. The same diameter as the desired thread.
    b. A few thousandths larger than the desired thread.
    c. A few thousandths (0.003"–0.004") smaller than the threads.
    d. All of the above.
    e. None of the above.

48. The drill used to make the hole prior to threading, is   48. _____
    called a _____.

49. How does the UNC thread series differ from the UNF thread series? _____

_____

6-10
(continued)

Name: _____

50. List the correct sequence taps should be used to form threads the full depth of a blind hole.

_____

51. Should a shaft be larger or smaller than the finished size if external threads are to be cut on it?

_____

52. Taps are turned in with a _____ . A _____ is used with dies.

52. _____

_____

53. What is an abrasive? _____

_____

54. _____ surfaces are never polished with an abrasive.

54. _____

Chapter **7**

Fasteners

LEARNING OBJECTIVES

After studying this chapter, students will be able to:
- ○ Identify several types of fasteners.
- ○ Explain why inch-based fasteners are not interchangeable with metric-based fasteners.
- ○ Describe how some fasteners are used.
- ○ Select the proper fastening technique for a specific job.
- ○ Describe chemical fastening techniques.

INSTRUCTIONAL MATERIALS

Text: pages 127–142
    Test Your Knowledge Questions, pages 140–141
Workbook: pages 47–50
Instructor's Resource: pages 121–130
    Guide for Lesson Planning
    Research and Development Ideas
    Reproducible Masters:
        7-1 Identifying Metric Fasteners
        7-2 Relative Strength of Hex Head Cap Screws
        7-3 Various Types of Cap Screws
        7-4 Test Your Knowledge Questions
    Color Transparencies (Binder/CD only)

GUIDE FOR LESSON PLANNING

This chapter serves to present the numerous types of fasteners that a machinist will use as well as their proper handling and installation. Although every type of fastener cannot be illustrated in the space available, the basic types are defined and illustrated.

The following items will aid in teaching this chapter:

- A selection of fasteners for examination.
- Examples showing how several types of fasteners are used.
- A selection of adhesives suitable for bonding metal.

Have students read and study Chapter 7, *Fasteners*. They should pay particular attention to the illustrations. Review the material and discuss the following:

- Why there are so many types of fasteners.
- How threaded fasteners are measured.
- Why there is a need for metric fasteners.
- How to identify metric fasteners.
- Thread nomenclature.
- The various fasteners available in the shop/lab.

Ask the following questions during the discussion:

- What unusual fasteners have students observed?
- Why is it not possible to substitute metric-based fasteners for inch-based fasteners?

- Where are metric fasteners used today?
- What problems are encountered when inch-based and metric-based fasteners are used on the same product? (Some automobiles use both types of fasteners.)
- Why are metric fasteners used today?
- Where would stainless steel fasteners be used?
- Why are the special fasteners used on aircraft so expensive?
- Is any student/trainee aware of a product where adhesives are used to join metal parts?
- A specially designed fastener is used to mount alloy wheels on some automobiles. Why are they used? How do they differ from conventional fasteners?

Technical Terms

Review the terms introduced in the chapter. New terms can be assigned as a quiz, homework, or extra credit. These terms are also listed at the beginning of the chapter.

> *adhesives*
> *assembly*
> *cyanoacrylate quick setting adhesives*
> *fastener*
> *keyway*
> *machine bolts*
> *permanent assemblies*
> *setscrews*
> *threaded fasteners*
> *washers*

Review Questions

Assign *Test Your Knowledge* questions. Copy and distribute Reproducible Master 7-4 or have students use the questions on pages 140–141 and write their answers on a separate sheet of paper.

Workbook Assignment

Assign Chapter 7 of the *Machining Fundamentals Workbook.*

Research and Development

Discuss the following topics in class or have students complete projects on their own.

1. Prepare samples of work using various threaded and nonthreaded fasteners. Mount them on a panel or arrange them on a table for the class to examine.

2. Secure samples of fasteners *not* described in this chapter. Classify them according to the material on which they are used and their recommended applications.

3. Develop a paper on how early fasteners were made. Use drawings to show how they looked.

4. Contact a manufacturer of fasteners and request samples of a machine bolt or cap screw in the various stages it must pass through until becoming a finished product. If it is not possible to secure actual samples, make a drawing showing the various stages of bolt manufacture.

5. Make a display of the various fasteners explained and described in this chapter. Mount and label them on a display panel.

6. Collect catalogs on fasteners for the school's technical library.

7. Secure samples of several adhesives suitable for joining metal to metal. Demonstrate the proper and safe way to use these materials.

8. Devise a test method that will determine the strength of the various adhesives.

9. Organize and label the storage of fasteners in your training area. Inventory them and determine which fasteners will have to be reordered.

TEST YOUR KNOWLEDGE ANSWERS, Pages 140–141

1. 1 1/2
2. Evaluate individually. Refer to Section 7.1.
3. Machine
4. Identification marks (inch size) and Class number (metric size) indicate bolt strength.
5. An application of penetrating oil.
6. setscrew
7. stud
8. Rivets, adhesives
9. To prevent nuts and/or bolts from vibrating loose.
10. retaining rings
11. To lock a full size nut in place.
12. wing
13. f. Used to make permanent assemblies.
14. h. Locks a regular nut in place.
15. e. Is hammered into a drilled or punched hole.

16. i. Eliminate costly tapping operations.
17. d. Protects projecting threads.
18. b. Used where parts must be aligned accurately and held in absolute relation with one another.
19. a. Developed for use in confined area, where a joint is only accessible from one side.
20. g. Slot cut in gear or pulley to receive "c."
21. j. Slot cut in shaft to receive "c."
22. c. Prevents a pulley or gear from slipping on a shaft.
23. Evaluate individually. Refer to Section 7.3.1.
24. Evaluate individually. Refer to Section 7.4.

WORKBOOK ANSWERS,
Pages 47–50

1. c. they permit work to be assembled and disassembled without damage to the parts
2. bolts
3. Cap
4. A. Clutch
   B. Cross Recess Type 1
   C. Cross Recess Type 2
   D. Flat
   E. Oval
   F. Fillister
   G. Truss
   H. Socket
   I. Slotted
   J. Round
   K. Pan
   L. Socket
5. d. All of the above.
6. square, hexagonal
7. jam, lock
8. washer
9. permanent
10. A. Internal-external
    B. Internal
    C. External
    D. Countersunk
    E. Split-ring
11. c. the joint is available from only one side
12. cotter pin
13. retaining, machining
14. They are hammered into a drilled or punched hole of the proper size.
15. When parts must be accurately positioned and held in absolute relation to one another.
16. key
17. keyseat, keyway
18. A. Surface preparation
    B. Adhesive preparation
    C. Adhesive application
    D. Assembly
    E. Bond development
19. Evaluate individually. Refer to Section 7.4.
20. A. Thread symbol for ISO (metric) thread
    B. Major diameter of threads in millimeters
    C. Pitch of threads in millimeters
    D. Thread tolerance class symbol (class of fit)
    E. Major diameter of thread in inches
    F. Threads per inch (pitch = 1/number of threads per inch)
    G. Thread series (Unified National Coarse)
    H. Class of fit (thread tolerance)

Identifying Metric Fasteners

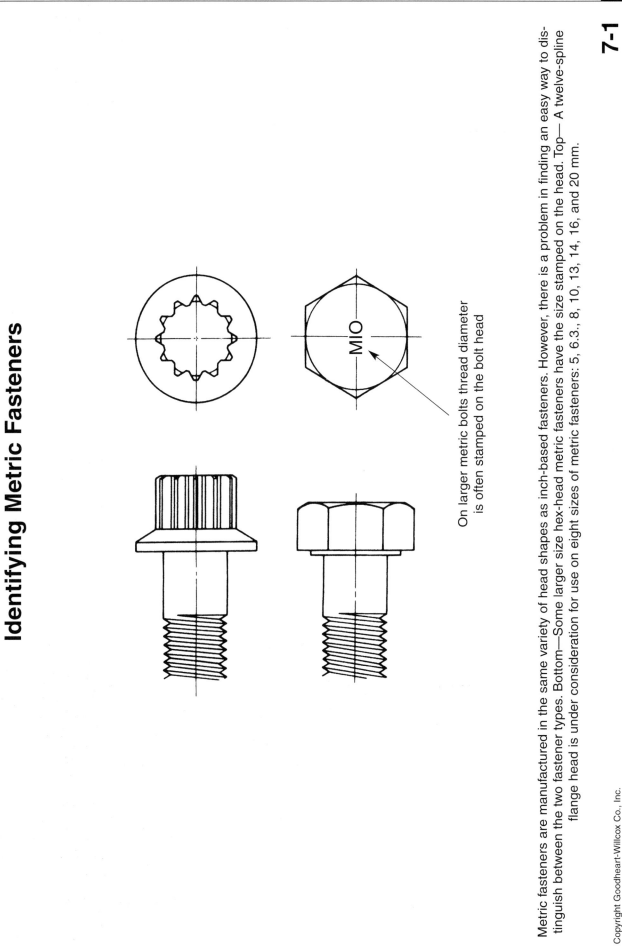

M10

On larger metric bolts thread diameter is often stamped on the bolt head

Metric fasteners are manufactured in the same variety of head shapes as inch-based fasteners. However, there is a problem in finding an easy way to distinguish between the two fastener types. Bottom—Some larger size hex-head metric fasteners have the size stamped on the head. Top— A twelve-spline flange head is under consideration for use on eight sizes of metric fasteners: 5, 6.3, 8, 10, 13, 14, 16, and 20 mm.

7-1

Relative Strength of Hex Head Cap Screws

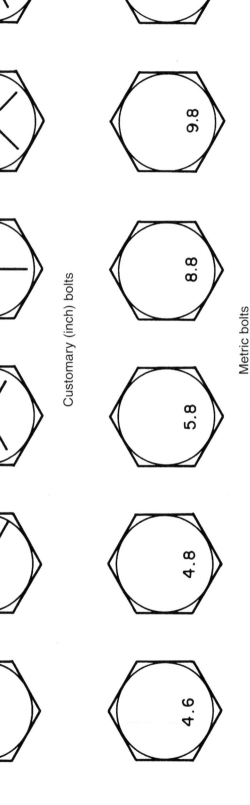

Customary (inch) bolts

Metric bolts

Identification marks (inch size) and class numbers (metric size) are used to indicate the relative strength of hex head cap screws. As identification marks increase in number, or class numbers become larger, increasing strength is indicated.

7-3

Various Types of Cap Screws

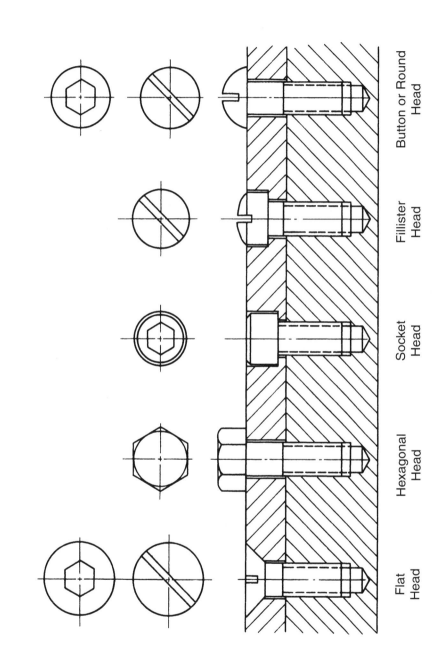

Flat Head

Hexagonal Head

Socket Head

Fillister Head

Button or Round Head

Fasteners

Name: _____ Date: _____ Score: _____

1. For maximum strength, a threaded fastener should screw   1. _____
   into its mating part a distance equal to _____ times the
   diameter of the thread.

2. There are many ways of joining material. List four types of threaded fasteners. Describe how
   each is used. _____

   _____

   _____

   _____

   _____

   _____

3. _____ screws are used for general assembly work.   3. _____

4. How is the strength of hex-head cap screws indicated? _____

   _____

5. When removing stubborn sheared bolts, what can be done to make their removal easier?

   _____

6. To prevent a pulley from slipping on a shaft, a _____ is   6. _____
   often employed.

7. The _____ bolt is threaded at both ends.   7. _____

8. _____ or _____ are employed when the parts are to be   8. _____
   joined permanently.
                                                           _____

9. Why are lock washers used? _____

   _____

10. While most _____ must be seated in grooves, a self-locking   10. _____
    type does not require the special recess.

11. When is a jam nut employed? _____

   _____

12. The shape of the _____ nut permits it to be loosened and   12. _____
    tightened without a wrench.

7-4

(continued)

Name: _____

- Match each word in the left column with the most correct sentence in the right column. Place the appropriate letter in the blank.

_____ 13. Rivet	a. Developed for use in confined area, where a joint is only accessible from one side.
_____ 14. Jam nut	b. Used where parts must be aligned accurately and held in absolute relation with one another.
_____ 15. Drive screw	
_____ 16. Thread-cutting screw	c. Prevents a pulley or gear from slipping on a shaft.
_____ 17. Acorn nut	d. Protects projecting threads.
_____ 18. Dowel pin	e. Is hammered into a drilled or punched hole.
_____ 19. Blind rivet	f. Used to make permanent assemblies.
_____ 20. Keyway	g. Slot cut in gear or pulley to receive "c."
_____ 21. Keyseat	h. Locks a regular nut in place.
_____ 22. Key	i. Eliminate costly tapping operations.
	j. Slot cut in shaft to receive "c."

23. List the steps, in their proper sequence, that must be used to join metals with adhesives.

_____

_____

_____

_____

_____

_____

_____

24. List at least five safety precautions that must be observed when using fasteners.

_____

_____

_____

_____

_____

_____

_____

7-4

Chapter 8

Jigs and Fixtures

LEARNING OBJECTIVES

After studying this chapter, students will be able to:
○ Explain why jigs and fixtures are used.
○ Describe a jig.
○ Describe a fixture.
○ Elaborate on the classifications of jigs and fixtures.

INSTRUCTIONAL MATERIALS

Text: pages 143–148
    Test Your Knowledge Questions, page 148
Workbook: pages 51–52
Instructor's Resource: pages 131–134
    Guide for Lesson Planning
    Research and Development Ideas
    Reproducible Master:
        8-1 Test Your Knowledge Questions
    Color Transparency (Binder/CD only)

GUIDE FOR LESSON PLANNING

This chapter is a brief introduction to the types and uses of jigs and fixtures in the machining industry. If possible, have a selection of simple jigs and fixtures available for examination with the parts produced using them.

Have students read and study Chapter 8, *Jig and Fixtures.* Review the assignment and discuss the following:
• How a jig differs from a fixture.
• Why they are used.
• The various types of jigs and fixtures.

Technical Terms

Review the terms introduced in the chapter.

New terms can be assigned as a quiz, homework, or extra credit. These terms are also listed at the beginning of the chapter.

> *box jig*
> *bushings*
> *closed jig*
> *drill template*
> *fixture*
> *fixture holding devices*
> *jig*
> *open jig*
> *plate jig*
> *slip bushings*

Review Questions

Assign *Test Your Knowledge* questions. Copy and distribute Reproducible Master 8-1 or have students use the questions on page 148 and write their answers on a separate sheet of paper.

Workbook Assignment

Assign Chapter 8 of the *Machining Fundamentals Workbook.*

Research and Development

Discuss the following topics in class or have students complete projects on their own.

1. Contact local industry and borrow examples of small jigs and fixtures they no longer use. Explain them to the class. If possible, include jobs produced on them.
2. Secure samples of products that have components produced with a jig or fixture.
3. Make a bulletin board display of magazine illustrations, drawings, and photographs showing various kinds of jigs and fixtures.
4. Design and manufacture a simple template jig for a job to be produced in the training area.
5. Design and manufacture a fixture for a training area product to be machined on a lathe, grinder, or drill press. Work in close cooperation with the drafting department in designing the product and producing the prints.
6. Seek permission to visit a machine shop using jigs and fixtures. Take 35 mm slides of the jigs and fixtures in use. Using the slides, give a presentation to the class.

TEST YOUR KNOWLEDGE ANSWERS, Page 148

1. production machine shops, hold work
2. They position the work and guide the cutting tool(s) so that all of the parts produced are uniform and within specifications. They are also used to hold work during assembly operations.

3. A jig is a device that holds work in place and guides the cutting tool during machining operations such as drilling, reaming, and tapping.
4. Open and box (closed) jigs. Evaluate description individually. Refer to Section 8.1.1.
5. several different operations must be performed on a job
6. A fixture is a device used to position and hold work while machining operations are performed. It does not guide the cutting tool(s).

WORKBOOK ANSWERS, Pages 51–52

1. b. usually nested between guide bars
2. d. All of the above.
3. manufacturing costs
4. b. encloses
5. c. in several directions
6. a. to guide the drills
7. a. hold work while machining operations are performed
8. d. All of the above.
9. holding
10. A. Jig
    B. Guide bars
    C. Drill press table

Jigs and Fixtures

Name: _____ Date: _____ Score: _____

1. Jigs and fixtures are devices used in _____ to _____ while   1. _____
   machining operations are performed.
   _____

2. Why are jigs and fixtures used? _____

   _____

   _____

   _____

   _____

3. What is a jig? _____

   _____

   _____

   _____

4. Jigs fall into two general types. List and briefly describe each type. _____

   _____

   _____

   _____

   _____

   _____

   _____

   _____

5. A combination of the two jig types listed in Question 4 is often used when _____.

   _____

6. What is a fixture? _____

   _____

   _____

   _____

8-1

Chapter 9

Cutting Fluids

LEARNING OBJECTIVES

After studying this chapter, students will be able to:
- ○ Understand why cutting fluids are necessary.
- ○ List the types of cutting fluids.
- ○ Describe each type of cutting fluid.
- ○ Discuss how cutting fluids should be applied.

INSTRUCTIONAL MATERIALS

Text: pages 149–152
    Test Your Knowledge Questions, page 152
Workbook: pages 53–54
Instructor's Resource: pages 135–138
    Guide for Lesson Planning
    Research and Development Ideas
    Reproducible Master:
        9-1 Test Your Knowledge Questions
    Color Transparency (Binder/CD only)

GUIDE FOR LESSON PLANNING

Have students read and study the chapter. Review the assignment with them and discuss the following:

- The function of cutting fluids.
- Types of cutting fluids.
- When each type of cutting fluid is used.
- How cutting fluids are applied.
- When certain types of cutting fluids should not be used.
- The need to wash hands thoroughly after using cutting fluids.

Technical Terms

Review the terms introduced in the chapter. New terms can be assigned as a quiz, homework, or extra credit. These terms are also listed at the beginning of the chapter.

chemical cutting fluids
contaminants
cutting fluids
emulsifiable oils
gaseous fluid
lubricating
mineral oils
misting
noncorrosive
semichemical cutting fluids

Review Questions

Assign *Test Your Knowledge* questions. Copy and distribute Reproducible Master 9-1 or have students use the questions on page 152 and write their answers on a separate sheet of paper.

Workbook Assignment

Assign Chapter 9 of the *Machining Fundamentals Workbook.*

Research And Development

Discuss the following topics in class or have students complete projects on their own.

1. Contact cutting fluid manufacturers for literature on their products.

2. Prepare charts on the cutting fluids recommended for use when machining various materials.

3. Inspect the coolant equipment on the machine tools in the shop/lab. Prepare a report on their condition and, if necessary, make recommendations on how they can be improved.

TEST YOUR KNOWLEDGE ANSWERS, Page 152

1. Answers will vary but may include the following: cool the work and cutting tool; improve surface finish quality; lubricating to reduce friction and cutting forces, thereby extending tool life; minimize material buildup on cutting tool edges; protect machines surfaces against corrosion; and flush away chips.

2. Any order: mineral oils, emulsifiable oils, chemical and semichemical fluids, gaseous fluids.

3. Mineral oil cutting fluid.

4. Answers will vary but may include the following: high cost, operator health problems, stain some metals, have a tendency to become rancid.

5. Emulsifiable oil

6. Answers will vary but may include the following: increased cooling capacity, cleaner to work with, less danger to health, and present no fire hazard.

7. Chemical

8. semichemical

9. Evaluate individually. Refer to Section 9.1.3.

10. Blows away chips at high velocity.

WORKBOOK ANSWERS, Pages 53–54

1. mineral

2. light-duty (low speed, light feed) operations where high levels of cooling and lubrication are not required

3. d. All of the above.

4. An approved respirator must be worn.

5. d. All of the above.

6. wetting

7. d. All of the above.

8. Some formulas have minimal lubricating qualities; they may cause skin irritation in some workers; and when they become contaminated with other oils, disposal can be problematic.

9. water-based

10. cooling, chips

11. cooling rates

Cutting Fluids

Name: _____ Date: _____ Score: _____

1. Cutting fluids must do many things simultaneously. What does this include? _____

   _____

   _____

   _____

   _____

2. List the four basic types of cutting fluids. _____

   _____

   _____

3. What type cutting oil is recommended for machining aluminum, magnesium, brass, and free-machining steels? _____

   _____

4. Why does the above type of cutting fluid have limited use? _____

   _____

   _____

5. _____ cutting fluids are also known as soluble oils.   5. _____

6. What advantages do the emulsifiable oil cutting fluids have over the cutting fluids indicated in Question 3? _____

   _____

7. _____ cutting fluids contain no oils.   7. _____

8. When small amounts of mineral oil are added to the   8. _____
   cutting fluid described in Question 7, it is known as
   _____ cutting fluid.

9. What are the advantages of the cutting fluids indicated in Questions 7 and 8? _____

   _____

   _____

   _____

10. What is dangerous about using compressed air to cool the area being machined? _____

    _____

    _____

    _____

9-1

Chapter 10

Drills and Drilling Machines

LEARNING OBJECTIVES

After studying this chapter, students will be able to:
- ○ Select and safely use the correct drills and drilling machine for a given job.
- ○ Make safe setups on a drill press.
- ○ Explain the safety rules that pertain to drilling operations.
- ○ List various drill series.
- ○ Sharpen a twist drill.

INSTRUCTIONAL MATERIALS

Text: pages 153–182
    Test Your Knowledge Questions, page 182
Workbook: pages 55–60
Instructor's Resource: pages 139–152
    Guide for Lesson Planning
    Research and Development Ideas
    Reproducible Masters:
        10-1 How a Drill Cuts
        10-2 Parts of a Twist Drill
        10-3 Clamping Work for Drilling
        10-4 Sharpening a Drill
        10-5 Centering Round Stock
        10-6 Counterbored Hole
        10-7 Spotfaced Hole
        10-8 Test Your Knowledge Questions
    Color Transparency (Binder/CD only)

GUIDE FOR LESSON PLANNING

This chapter introduces various types of drills and drilling machines and explains basic drilling practices. An assortment of drilling equipment (drills, drill gage, center finder, center drill, sleeve, socket, drift, vises, parallels), should be available for student examination.

Have students/trainees read and study the chapter. Review the assignment and discuss and demonstrate the following:

- Definition of a machine tool.
- Types of drilling machines.
- Variety of drill press machining operations.
- How drill press size is determined.
- How a twist drill cuts.
- Why tool is called a twist drill.
- Types of drills and drill sizes.
- Ways to determine drill size.
- Parts of a drill.
- How drills can be mounted in a drill press.
- Work-holding devices and setups.
- Cutting speeds and feeds and their importance.
- Using a center finder to position drill.
- Proper sequence for drilling a hole.
- Cutting fluids and when they should or should not be used.

- Reason for pilot hole and determining its size.
- Holding and centering round stock for drilling.
- Reamers and reaming.
- Countersinking, counterboring, and spot-facing.
- Safety procedures to be observed when using drilling machines.

Emphasize drilling safety, especially the importance of mounting work solidly to the work table to prevent the dangerous "merry-go-round."

Before demonstrating drill press operations, be sure the tools and equipment are in safe operating condition with all guards and safety devices in place. Students must wear approved eye protection while observing the demonstrations.

Briefly review the demonstrations and encourage students to ask questions.

Technical Terms

Review the terms introduced in the chapter. New terms can be assigned as a quiz, homework, or extra credit. The following terms are also listed at the beginning of the chapter.

> *blind hole*
> *center finder*
> *countersinking*
> *drill point gage*
> *flutes*
> *lip clearance*
> *machine reamer*
> *multiple spindle drilling machines*
> *spotfacing*
> *twist drills*

Review Questions

Assign *Test Your Knowledge* questions. Copy and distribute Reproducible Master 10-8 or have students use the questions on page 182 and write their answers on a separate sheet of paper.

Workbook Assignment

Assign Chapter 10 of the *Machining Fundamentals Workbook.*

Research and Development

Discuss the following topics in class or have students complete projects on their own.
1. Drills are expensive. Each semester, keep a record of drills broken in the shop and the cause of breakage. Make recommendations for reducing drill breakage and damage.
2. Make a series of safety posters on the use of a drill press.
3. Prepare a research paper on early drilling devices. Include sketches. You may want to reproduce this report or make a series of transparencies for the overhead projector.
4. Develop a research project investigating the effects of cutting fluids and compounds on drilling (quality of finished hole, etc.). Include samples of holes drilled in the same material with and without coolant/cutting fluid.
5. Prepare a teaching aid that will show examples of a drilled hole, reamed hole, countersinking, spotfacing, and counterboring.
6. Borrow drill jigs from a local industry. Explain how they are used.
7. Demonstrate one of the following:
   - Centering round stock in a V-block.
   - The proper way to use a wiggler.
   - Sharpening a twist drill.
   - Several methods of safely clamping work on a drill press table.

TEST YOUR KNOWLEDGE ANSWERS, Page 182

1. c. Rotating against material with sufficient pressure to cause penetration.
2. It is determined by the largest diameter of a circular piece that can be drilled on center.
3. c. Both of the above.
4. Fractional, Number, Letter, and Metric.
5. By micrometer and drill gage.
6. Straight shank and taper shank.
7. Straight
8. Taper
9. flutes
10. d. All of the above.
11. Sleeve
12. socket
13. Drift
14. d. All of the above.
15. Lip clearance, length and angle of lips, and proper location of dead center.
16. One lip will cut and hole will be oversized and out-of-round.
17. drill point gage

18. 118
19. Compressed air
20. pilot or lead, dead center
21. A hole that does not go all of the way through the work.
22. The distance the full diameter goes into the work.
23. jobber's reamer (machine reamer)
24. solid
25. It should be removed before stopping the machine.
26. two-thirds
27. Countersinking
28. counterboring
29. Spotfacing

WORKBOOK ANSWERS, Pages 55–60

1. drill press
2. b. the largest diameter circular piece that can be drilled on center
3. c. Bench drill presses
4. a. Radial drill presses
5. d. All of the above.
6. d. All of the above.
7. high-speed (HSS), carbon
8. titanium nitride
9. b. oil-hole drill
10. e. None of the above.
11. drill margins
12. b. Dead center
13. a. provides a means of separating the taper from the holding device
14. d. All of the above.
15. d. Either a or b.
16. Parallels
17. angular
18. d. All of the above.
19. U-strap
20. A. Sleeve
    B. Socket
    C. Drift
21. enlarge smaller taper shank to fit drill press spindle
22. reduce taper shank so it will fit drill press spindle
23. separate taper shank from sleeve, socket, or drill from drill press spindle
24. Check with drill point gage.
25. a. distance that the drill cutting edge circumference travels per minute
26. does not cause
27. 2400 rpm
28. 4000 rpm
29. 480 rpm
30. 680 rpm
31. drilling a hole in soft metal and observing chip formation; When properly sharpened, chips will come out of the flutes in curled spirals of equal size and length.
32. d. All of the above.
33. V-blocks
34. To receive flat-headed fasteners.
35. The operation machines a flat circular area on a rough surface to provide a bearing surface for the head of a bolt, washer, or nut.
36. d. All of the above.
37. mounts on a special arbor that can be used with several reamer sizes
38. b. rose chucking
39. Evaluate individually.
40. Evaluate individually.

How a Drill Cuts

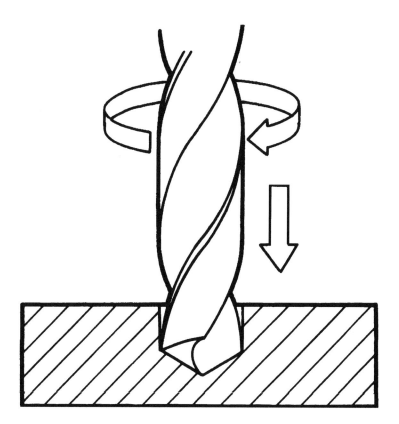

Drilling is the operation most often performed on a drill press. Both rotating force and a downward pushing force are needed for drilling.

10-1

Parts of a Twist Drill

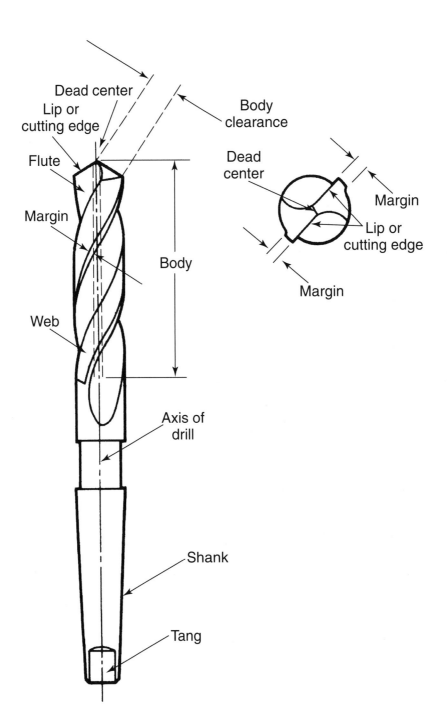

10-2

Clamping Work for Drilling

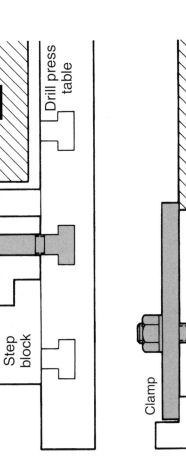

Correct clamping technique. Note that clamp is parallel to work. Clamp slippage can be reduced by placing a piece of paper between the work and the clamp.

Incorrect clamping technique. T-bolt is too far from work. This allows the clamp to spring under pressure.

Sharpening a Drill

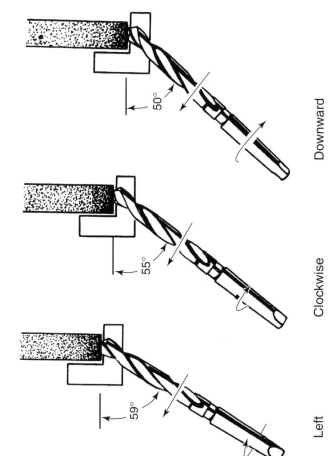

50°

Downward

55°

Clockwise

59°

Left

Hold the point lightly against the rotating wheel and use three motions of the shank: to the left, clockwise rotation, and downward.

Centering Round Stock

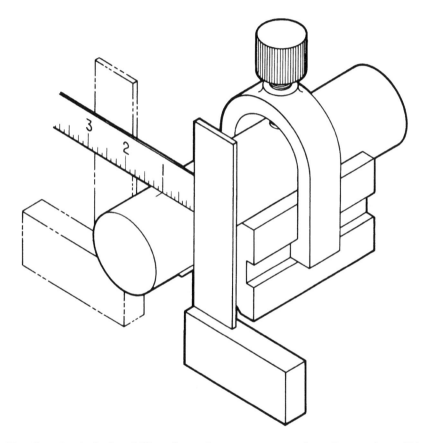

To align the hole for drilling through exact center, place the work and V-block on the drill press table or on a surface plate. Rotate the punch mark until it is upright. Place a steel square on the flat surface with the blade against the round stock as shown above. Measure from the square blade to the punch mark, and rotate the stock until the measurement is the same when taken from both sides of the stock.

10-5

Counterbored Hole

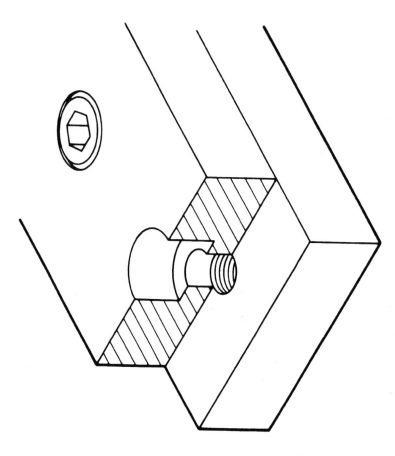

A sectional view of a hole that has been drilled and counterbored to receive a socket-head screw.

Spotfaced Hole

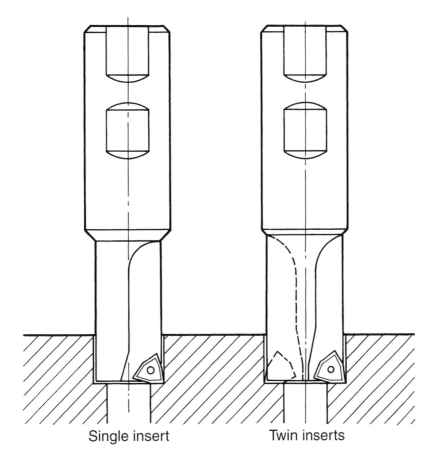

Single insert   Twin inserts

Counterbores with carbide indexable inserts. The inserts are rotated
when a cutting edge becomes dull.

Drills and Drilling Machines

Name: _____ Date: _____ Score: _____

1. A twist drill works by:
   a. being forced into material.
   b. rotating against material and being pulled through by the spiral flutes.
   c. Rotating against material with sufficient pressure to cause penetration.
   d. All of the above.
   e. None of the above.

1. _____

2. How is drill press size determined? _____

   _____

3. Drills are made from:
   a. high-speed steel.
   b. carbon steel.
   c. Both of the above.
   d. Neither of the above.

3. _____

4. Drill sizes are expressed by what four series? _____

   _____

5. What are two techniques used to determine a drill's size? _____

   _____

6. List the two types of drill shanks. _____

   _____

7. _____ shank drills are used with a chuck.

7. _____

8. _____ shank drills fit directly into the drill press spindle.

8. _____

9. The spiral grooves that run the length of the drill body are called _____.

9. _____

10. The spiral grooves in a drill body are used to:
    a. help form the cutting edge of the drill point.
    b. curl chips for easier removal.
    c. form channels through which the chips can escape from the hole.
    d. All of the above.
    e. None of the above.

10. _____

11. Name the device employed to enlarge a taper shank drill so it will fit the spindle opening.

11. _____

12. The device used to permit a drill with a taper shank too large to fit the spindle opening is called a _____.

12. _____

13. What is the name of the tool used to separate a taper shank drill from the above devices in question # 12?

13. _____

10-8

(continued)

Name: _____

14. Cutting fluids or compounds are used to:
    a. cool the drill.
    b. improve the finish of a drilled hole.
    c. Aid in the removal of chips.
    d. All of the above.
    e. None of the above.

14. _____

15. List the three factors that must be considered when repointing a drill. _____

_____

16. What occurs when the cutting lips of a drill are not sharpened to the same lengths? _____

_____

17. The _____ should be used frequently when sharpening to ensure a correctly sharpened drill.

17. _____

18. The included angle of a drill point sharpened for general drilling is _____ degrees.

18. _____

19. What coolant should be used when drilling cast iron?

19. _____

20. Large drills require a considerable amount of power and pressure to get started. They also have a tendency to drift off center. These conditions can be minimized by first drilling a _____ hole. This hole should be as large as, or slightly larger than, the width of the _____ of the drill point.

20. _____

_____

21. What is a blind hole? _____

_____

22. How is the depth of a drilled hole measured? _____

_____

23. The _____ is almost identical to the hand reamer except that the shank has been designed for machine use.

23. _____

24. A _____ expansion reamer provides rigidity and accuracy not possible with conventional expansion reamers.

24. _____

25. How should a reamer be removed from a finished hole? _____

_____

26. The cutting speed for a high-speed reamer is approximately _____ that for a similar-sized drill.

26. _____

27. What is the name of the operation employed to cut a chamfer in a hole to receive a flat-head screw?

27. _____

28. The operation used to prepare a hole for a fillister or socket-head screw is called _____.

28. _____

29. _____ is the operation that machines a circular spot on a rough surface for the head of a bolt or nut.

29. _____

10-8

Chapter 11

Offhand Grinding

LEARNING OBJECTIVES

After studying this chapter, students will be able to:
- ○ Identify the various types of offhand grinders.
- ○ Dress and true a grinding wheel.
- ○ Prepare a grinder for safe operation.
- ○ Use an offhand grinder safely.
- ○ List safety rules for offhand grinding.

INSTRUCTIONAL MATERIALS

Text: pages 183–190
    Test Your Knowledge Questions, pages 189–190
Workbook: pages 61–64
Instructor's Resource: pages 153–160
    Guide for Lesson Planning
    Research and Development Ideas
    Reproducible Masters:
        11-1 Grinding Machine Operation
        11-2 Adjusting Grinder Tool Rest
        11-3 Using Wheel Dressers
        11-4 Test Your Knowledge Questions
    Color Transparency (Binder/CD only)

GUIDE FOR LESSON PLANNING

Have students read and study Chapter 11, paying attention to the illustrations. Reproducible Masters can be used on the overhead projector or copied and distributed to the class. Review the assignment and discuss or demonstrate the following:

- Definition of grinding. Use Reproducible Master 11-1.
- When offhand grinding is usually used.
- Types of offhand grinders.
- Checking grinding wheels for safe operation.
- Importance of properly adjusted tool rest. Use Reproducible Master 11-2.
- How to dress a grinding wheel. Use Reproducible Master 11-3.
- When to use a dry-type grinder.
- When a wet-type grinder is used.
- Safety procedures to be observed when using offhand grinding machines.

When demonstrating offhand grinding techniques, be sure that:

- Equipment is properly adjusted with all guards and safety devices in place.
- Students/trainees are wearing approved eye protection.
- Grinding wheels are solid, dressed, and running true.
- Students understand what metals can or cannot be ground on shop/lab grinding machines.

Briefly review the demonstrations and encourage students to ask questions.

Technical Terms

Review the terms introduced in the chapter. New terms can be assigned as homework, extra credit, or used for a quiz. The following list is also given at the beginning of the chapter.

abrasive belt   precision microgrinder
   grinding machines   reciprocating
bench grinder   hand grinder
concentricity   temper
flexible shaft grinders   tool rest
pedestal grinder   wheel dresser

Review Questions

Assign *Test Your Knowledge* questions. Copy and distribute Reproducible Master 11-4 or have students use the questions on pages 189–190 and write their answers on a separate sheet of paper.

Workbook Assignment

Assign Chapter 11 of the *Machining Fundamentals Workbook.*

Research and Development

Discuss the following topics in class or have students complete projects on their own.

1. How are natural sandstone grinding wheels made? Secure samples of natural sandstone and compare them with manufactured abrasives. What, in your opinion, makes the manufactured abrasive superior to the natural product?
2. Research the term *MOH SCALE.* Prepare a chart showing the common abrasives in order of their hardness. Secure samples to mount on the chart.
3. Grinders used by 18th century workers required them to work in an unusual position. Prepare a brief presentation for the class on how they worked, some of the health problems they had because of their job, and some of the articles they produced. Create visual aids (photos, drawings, transparencies), for use during the presentation.
4. Borrow an old fashioned natural sandstone grinding machine. Demonstrate its use to the class.
5. True and dress the grinding wheels used on the machines in the shop.
6. Secure a list of the various grades and types of grinding wheels available from a typical abrasive manufacturer.

TEST YOUR KNOWLEDGE ANSWERS, Pages 189–190

1. Grinding is an operation that removes material by rotating an abrasive wheel or belt against work.
2. Evaluate individually.
3. offhand
4. c. Work is manipulated with fingers until desired shape is obtained.
5. Dry type and wet type. The dry type does not use a flow of coolant on wheel. The wet type uses a flow of coolant over wheel.
6. 1/16, 1.5, pulled, rest, wheel
7. Supporting it on a section of wire and lightly tapping it with a light metal bar or screwdriver handle. A good wheel will give off a clear ringing sound.
8. c. Stand to one side of the grinder when using the machine.
9. burn
10. wet
11. crowned; It will minimize the amount of contact between the wheel and the work. This reduces the risk of the work being damaged or destroyed by excessive heat.
12. checking it for soundness
13. Evaluate individually. Refer to Section 11.4.

WORKBOOK ANSWERS, Pages 61–64

1. belt, disc
2. d. All of the above.
3. roughing, finish grinding
4. c. keeps the wheels constantly flooded with fluid
5. d. All of the above.
6. Serious injury could result if the cloth is pulled into the wheel.
7. d. All of the above.
8. wheel dresser
9. c. use the face of the wheel, *not* the sides
10. grooves/ridges
11. a. Diamond-impregnated
12. c. being damaged or destroyed by excessive heat buildup
13. greater
14. d. All of the above.
15. Evaluate individually. Refer to Section 11.4.

Grinding Machine Operation

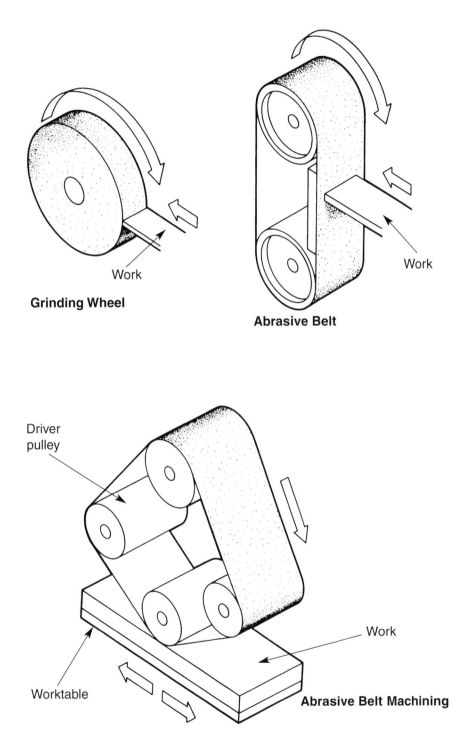

Grinding Wheel

Work

Abrasive Belt

Work

Driver pulley

Work

Worktable

Abrasive Belt Machining

11-1

Adjusting Grinder Tool Rest

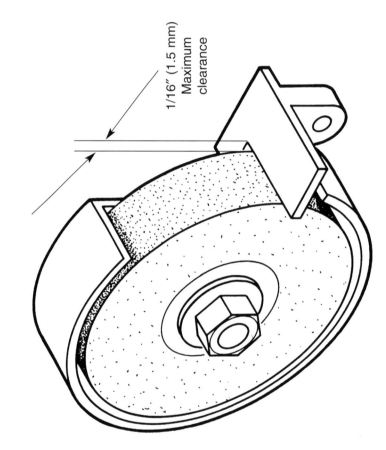

1/16″ (1.5 mm) Maximum clearance

Using Wheel Dressers

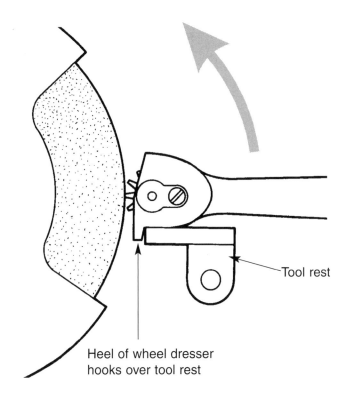

Tool rest

Heel of wheel dresser
hooks over tool rest

Move the tool back and forth
over the face of the stone.

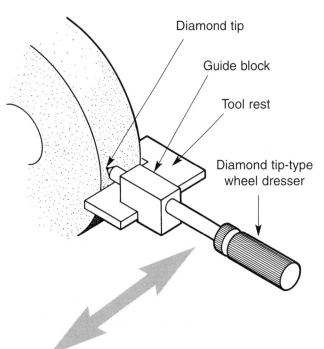

Diamond tip

Guide block

Tool rest

Diamond tip-type
wheel dresser

Industrial diamonds are also
used to dress and true grinding
wheels. The guide block is used
for grinders with slotted tool
rests.

11-3

Offhand Grinding

Name: _____ Date: _____ Score: _____

1. Describe the grinding operation. _____
   _____
   _____
   _____
   _____

2. How do abrasive belt grinders differ from abrasive wheel grinders? _____
   _____
   _____
   _____
   _____

3. Bench and pedestal grinders are used to do _____ grinding.

   3. _____

4. The grinding technique referred to in the preceding statement is so named because:
   a. It can only do external work.
   b. Work is too hard to be machined by other methods.
   c. Work is manipulated with fingers until desired shape is obtained.
   d. All of the above.
   e. None of the above.

   4. _____

5. Name the two types of pedestal grinders. How do they differ? _____
   _____
   _____
   _____

6. The tool rest should be about _____ inches or _____ mm away from the grinding wheel or belt for safety. This prevents the possibility of work being _____ between the tool _____ and _____.

   6. _____
   _____
   _____
   _____
   _____

Name: _____

7. How can grinding wheel soundness be checked? _____

   _____

   _____

8. Since a grinding wheel cannot be checked each time the   8. _____
   grinder is used, it is recommended that the operator:
   a. Not use the grinder.
   b. Check with the instructor whether the wheel is sound.
   c. Stand to one side of the grinder when using the machine.
   d. All of the above.
   e. None of the above.

9. Work will _____ if it is forced against the wheel with too   9. _____
   much pressure.

10. Carbide-tipped tools are usually sharpened on a _____   10. _____
    grinder.

11. The face of wheel on a wet-type grinder is _____ slightly. Why is this done?_____

    _____

    _____

    _____

    _____

12. Never mount a grinding wheel on a grinder without   12. _____
    _____.

13. List four safety precautions to be observed when operating a grinder.

    _____

    _____

    _____

    _____

    _____

    _____

    _____

    _____

Chapter 12

Sawing and Cutoff Machines

INSTRUCTIONAL MATERIALS

Text: pages 191–200
    Test Your Knowledge Questions, page 199
Workbook: pages 65–68
Instructor's Resource: pages 161–170
    Guide for Lesson Planning
    Research and Development Ideas
    Reproducible Masters:
        12-1 Cutoff Saws
        12-2 Cutting Pressure
        12-3 Tooth Set and Tooth Shape
        12-4 Reverse Work after Replacing Blade
        12-5 Holding Work for Sawing
        12-6 Test Your Knowledge Questions
    Color Transparencies (Binder/CD only)

GUIDE FOR LESSON PLANNING

Have students read and study the chapter. Reproducible Masters can be used on the overhead projector or copied and distributed to the class. Review the assignment and discuss the following:
- Types of power saws and how they operate.
- Proper way to mount, position, and cut material.

- How to select the proper blade.
- How to mount and tension blades.
- How to adjust machine for most efficient cutting.
- Causes and corrections for sawing problems.
- Safety precautions to be observed when power sawing.

When demonstrating power saw operations be sure all students can hear and observe what you are doing. Approved eye protection must be worn by all students/trainees.

Briefly review the demonstrations. Provide students/trainees the opportunity to ask questions.

Technical Terms

Review the terms introduced in the chapter. New terms can be assigned as a quiz, homework, or extra credit. The following list is also given at the beginning of the chapter.
all-hard blade
cold circular saw
dry abrasive cutting
flexible-back blades

friction saw
gravity feed
horizontal band saw
raker set
three-tooth rule
wet abrasive cutting

Review Questions

Assign *Test Your Knowledge* questions. Copy and distribute Reproducible Master 12-6 or have students use the questions on page 199 and write their answers on a separate sheet of paper.

Workbook Assignment

Assign Chapter 12 of the *Machining Fundamentals Workbook.*

Research and Development

Discuss the following topics in class or have students complete projects on their own.

1. Make a display panel that includes samples of the different band saw blades used by industry. Label them according to their recommended use. Provide a magnifying glass so the blade teeth can be examined in detail.

2. Secure samples of abrasive cutoff wheels and of the material they are best suited to cut. Prepare a bulletin board display using these wheels and samples, and manufacturers' brochures of abrasive cutoff machines.

3. Prepare large scale models of the following blade types from suitable aluminum sheet: raker and wavy set teeth, standard tooth blade, skip tooth blade, and hook tooth blade. Mount them in hardwood stands for easy examination.

4. Design and produce a series of safety panels pertaining to power sawing.

5. If the power saw in your shop has seen extensive service, use the machine's service manual to overhaul the machine. If needed, contact the manufacturer for a service manual and parts list. If time permits, paint the machine according to "color dynamics" specifications.

6. Demonstrate the various methods used to properly tension a blade on a power saw.

7. Contact a saw blade manufacturer and request a chart showing the best type blade for various sawing situations. Mount it near the saw so it is readily accessible.

TEST YOUR KNOWLEDGE ANSWERS, Page 199

1. Reciprocating, continuous band-type, and circular-type.
2. reciprocating, back
3. A minimum of three teeth must be cutting at all times.
4. Cast iron, and soft brass
5. Flexible back, and all-hard.
6. a. 4
   b. 6
   c. 10- and 14-
7. Sound of blade when properly tensioned (low musical ring); shape of pin hole (slightly elongated); and using a torque wrench to tighten to manufacturer's specifications.
8. When several pieces of same length must be cut.
9. Faster speed, greater precision, and less waste.
10. Raker and wavy.
11. a. standard
    b. skip
    c. hook
12. Abrasive cutoff saw, cold circular saw, and friction saw.
13. Evaluate individually. Refer to Section 12.6.

WORKBOOK ANSWERS, Pages 65–68

1. c. definite pressure
2. b. coarse-tooth blades
3. c. Flexible-back
4. A. Raker
   B. Wavy
5. Checking wheel alignment, guide alignment, feed pressure, hydraulic systems.
6. A. Blade dropped on work. Loose blade or excessive speed.
   B. Usually caused by worn blade.
   C. Dirty mounting plates or too much tension on blade.
   D. Insufficient pressure/excessive pressure. Lack of coolant or poorly adjusted machine.
   E. Starting cut on sharp corner. Less than three teeth cutting, blade too fine or too coarse.
7. Wet, Dry
8. accurate, sever
9. billets
10. Evaluate individually.

12-1

Cutoff Saws

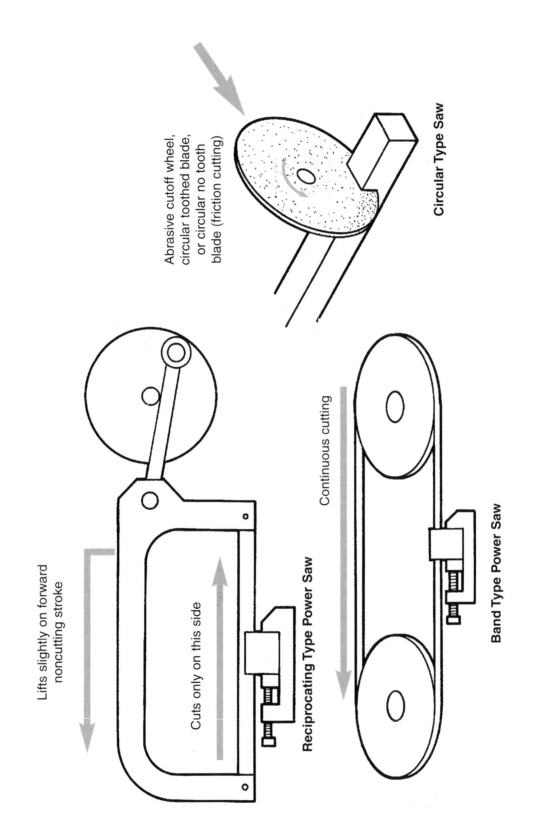

Abrasive cutoff wheel, circular toothed blade, or circular no tooth blade (friction cutting)

Circular Type Saw

Lifts slightly on forward noncutting stroke

Cuts only on this side

Reciprocating Type Power Saw

Continuous cutting

Band Type Power Saw

Cutting Pressure

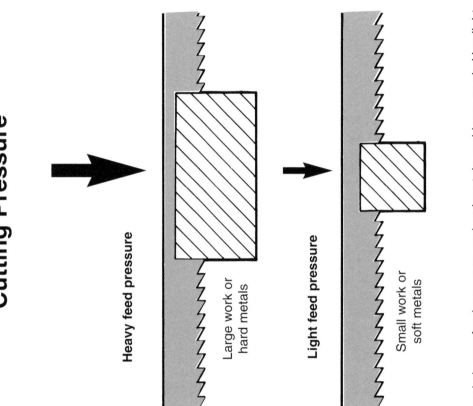

Heavy feed pressure

Large work or hard metals

Light feed pressure

Small work or soft metals

Apply heavy feed pressure on hard metals and large work. Use light pressure on soft metals and work with small cross sections.

12-3

Tooth Set and Tooth Shape

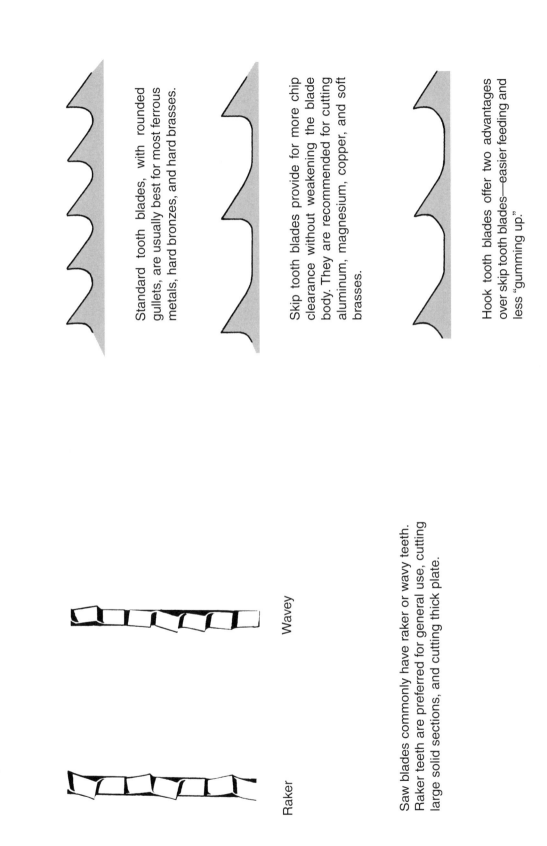

Standard tooth blades, with rounded gullets, are usually best for most ferrous metals, hard bronzes, and hard brasses.

Skip tooth blades provide for more chip clearance without weakening the blade body. They are recommended for cutting aluminum, magnesium, copper, and soft brasses.

Hook tooth blades offer two advantages over skip tooth blades—easier feeding and less "gumming up."

Wavey

Raker

Saw blades commonly have raker or wavy teeth. Raker teeth are preferred for general use, cutting large solid sections, and cutting thick plate.

Reverse Work after Replacing Blade

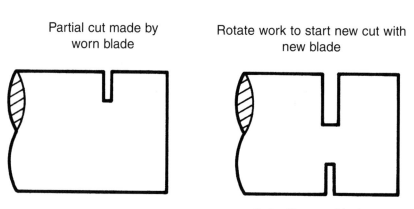

Partial cut made by worn blade

Rotate work to start new cut with new blade

Cut with worn blade

Never attempt to start a new blade in a cut made by a worn blade. Reverse the work and start another cut on the opposite side. Cut through to the old cut.

12-5

Holding Work for Sawing

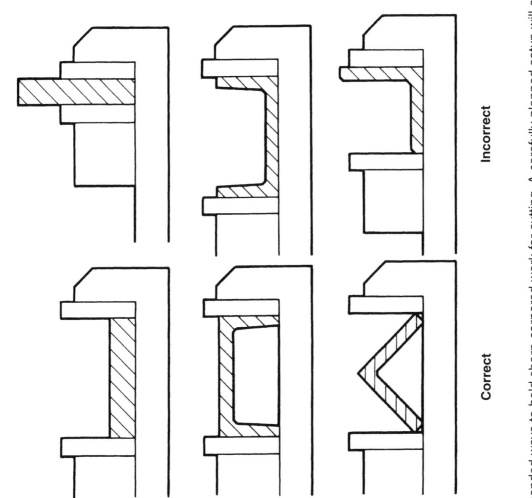

Correct

Incorrect

Recommended ways to hold sharp-cornered work for cutting. A carefully planned setup will ensure that at least three teeth will be cutting, greatly extending blade life

Sawing and Cutoff Machines

Name: _____ Date: _____ Score: _____

1. List the three basic types of metal-cutting saws. _____

_____

2. The _____ type saw has a back-and-forth cutting action. However, it only cuts on the _____ stroke.

2. _____

_____

3. What is the "three-tooth rule" for sawing? _____

_____

_____

_____

4. When using a power sawing machine, with which materials should you *not* use coolant?

_____

_____

5. Hacksaw blades are manufactured in two principal types. Name them.

5. _____

_____

6. The following "rule-of-thumb" should be followed for selecting the correct blade:

a. _____ teeth per inch for cutting large sections or readily machined materials.

b. _____ teeth per inch for cutting harder alloys and miscellaneous cutting.

c. _____ teeth per inch for cutting on the majority of light-duty machines, where work is limited to small sections and moderate to light feed pressures.

6. a. _____

b. _____

c. _____

7. List three methods used to put proper tension on a power hacksaw blade.

_____

_____

_____

_____

8. When is a stop gage used? _____

_____

_____

_____

Name: _____

9. What three advantages does the continuous band sawing machine offer over other types of power saws? _____

_____

_____

_____

10. Band saw blades are made with two types of teeth. Name them.

10. _____

_____

11. The tooth pattern of a blade determines the efficiency of a blade in various materials.
   a. The _____ tooth is best suited for cutting most ferrous metals.
   b. The _____ tooth pattern is preferred for cutting aluminum, magnesium, copper, and soft brass.
   c. The _____ tooth is also recommended for most nonferrous metallic materials.

11. a. _____

   b. _____

   c. _____

12. List the three types of circular metal-cutting saws. _____

_____

_____

13. List five safety precautions to be observed when operating a power saw.

_____

_____

_____

_____

_____

_____

_____

_____

_____

Chapter 13

The Lathe

INSTRUCTIONAL MATERIALS

Text: pages 201–240
   Test Your Knowledge Questions,
   pages 239–240
Workbook: pages 69–76
Instructor's Resource: pages 171–190
   Guide for Lesson Planning
   Research and Development Ideas
   Reproducible Masters:
      13-1 Lathe Operation
      13-2 Lathe Measurement
      13-3 Parts of a Lathe
      13-4 High-Speed Steel Cutting Tools
         (*nomenclature and shapes*)
      13-5 Sharpening HSS Cutter Bits
      13-6 Using the Cutter Bit Gage
      13-7 Calculating Cutting Speeds
      13-8 Cutting Speed and Feed Problems
      13-9 A, Checking Center Alignment
      13-9 B, Checking Center Alignment
      13-10 Facing in a Chuck
      13-11 Test Your Knowledge Questions
   Color Transparencies (Binder/CD only)

GUIDE FOR LESSON PLANNING

Because this chapter is rather extensive, it should be divided into several segments. Teach the segments that best suit your program.

Part I—Parts of the Lathe

Have students read and study pages 201–211. Review the assignment using Reproducible Masters 13-1, 13-2, and 13-3 as overhead transparencies and/or handouts. Discuss the following:
- How a lathe operates.
- How lathe size is determined.
- Major parts of the lathe.
- Preparing a lathe for operation.
- Cleaning a lathe.
- Lathe safety.
- Emphasize the importance of lubricating and checking over a lathe before operating.

Part II—Cutting Tools and Tool Holders

A selection of cutting tools and tool holders should be available for the class to examine.

Have students read and study pages 211–219. Review the assignment using Reproducible Masters 13-4, 13-5, and 13-6 as overhead transparencies and/or handouts. Discuss the following:
- High-speed steel (HSS) cutting tools and how they are shaped for different types of turning.

- How to sharpen high-speed steel cutting tools.
- Carbide-tipped cutting tools.
- Indexable insert cutting tools.
- How the shape of an insert determines its strength.
- The reason for a chip breaker on a single point tool.
- The nine basic categories of cutting tools.
- Emphasize how to handle sharpened cutting tools to prevent injury and premature dulling.

Part III—Cutting Speeds and Feeds

Have students read and study pages 220–222. Review the assignment using Reproducible Master 13-7 as an overhead transparency and/or handout. Discuss the following:

- The factors that effect cutting speeds and feeds.
- How to calculate cutting speeds and feeds. Use Reproducible Master 13-8 to provide practice in calculating cutting speeds and feeds.
- How lathe speed and carriage feed is set on the lathes in your shop/lab.
- Reason for making roughening and finishing cuts.
- How depth of cut is determined on lathes in your shop/lab.
- Demonstrate the difference between roughing cuts and finishing cuts. All students must wear approved eye protection during the demonstration.

Part IV—Work-Holding Attachments

Several lathes should be set up to show work mounted between centers, mounted in various types of chucks and collets, and bolted to a faceplate.

Have students read and study page 222. Discuss and demonstrate the various work holding attachments set up on the lathes. Explain the safety precautions that must be observed when mounting the attachments on the lathe and when they are being used.

Part V—Turning Between Centers

Have a lathe set up for turning between centers plus a selection of the equipment necessary for turning between centers.

Students should read and study pages 223–231. Review the assignment using Reproducible Masters 13-9 A and B as overhead transparencies and/or handouts. Demonstrate turning between centers. After the demonstration, discuss and encourage questions on the following:

- How to set up a lathe for turning between centers.
- Proper depth to drill center holes.
- How to check for center alignment.
- Selecting the proper size lathe dog.
- Proper way to mount work between centers. (Why a ball bearing center is preferred to a dead center.)
- Facing work mounted between centers.
- Facing to length.
- How to position the tool holder and cutting tool.
- Rough and finish turning.
- Turning to a shoulder.
- Grooving or necking operations.
- Emphasize the safety precautions that must be observed when turning between centers.

Part VI—Using Lathe Chucks

Set up lathes with the various types of chucks for demonstrations and student/trainee examination. Have students read and study pages 231–237.

Review the assignment using Reproducible Master 13-10 as an overhead transparency and/or handout. Demonstrate how the various types of chucks are used. Discuss the following:

- Advantages and disadvantages of the 3-jaw universal chuck.
- How to install jaws in the universal chuck.
- Advantages and disadvantages of the 4-jaw independent chuck.
- How to center work in an independent chuck.
- Using the Jacobs chuck in the tailstock and headstock.
- Advantages and disadvantages of the collet chuck.
- How to mount and remove chucks safely.
- Facing stock in a chuck. (How to tell whether the cutting tool is above or below center.)

- Plain turning and turning a shoulder.
- How to safely perform parting operations.
- Emphasize the safety precautions that must be observed when turning work mounted in a chuck.

Briefly review the demonstrations. Provide students with the opportunity to ask questions.

Technical Terms

Review the terms introduced in the chapter. New terms can be assigned as a quiz, homework, or extra credit. The following list is also given at the beginning of the chapter.

  compound rest
  cross-slide
  depth of cut
  facing
  headstock
  indexable insert cutting tools
  plain turning
  single-point cutting tool
  tailstock
  tool post

Review Questions

Assign *Test Your Knowledge* questions. Copy and distribute Reproducible Master 13-11 or have students use the questions on pages 239–240 and write their answers on a separate sheet of paper.

Workbook Assignment

Assign Chapter 13 of the *Machining Fundamentals Workbook.*

Research and Development

Discuss the following topics in class or have students complete projects on their own.

1. Make large scale wooden models of the basic cutting tool shapes. They should be cutaway models to permit the various clearance angles to be easily observed.
2. Prepare a comparison test using carbon steel, high-speed steel, and cemented carbide cutting tools. Make the tests on mild steel (annealed), tool steel (heat treated), and aluminum alloy. Employ the recommended cutting speeds and feeds. Make a graph that will show the times needed by the various cutting tools to perform an identical machining operation. Also indicate surface finish quality.
3. Develop and produce a series of posters on lathe safety.
4. Develop a research project to investigate the effects of cutting fluids upon the quality of the surface finish of turned work. Prepare a paper on your findings.
5. Show a film or video tape on the operation of a CNC lathe or turning center.

TEST YOUR KNOWLEDGE ANSWERS, Pages 239–240

1. c. The work rotating against the cutting tool, which is controllable.
2. swing, length, bed
3. c. The length of the bed minus the space taken up by the headstock and the tailstock.
4. d. All of the above.
5. They provide precise alignment of headstock and tailstock and serve as rails to guide the carriage.
6. tool travel, spindle revolution
7. a. Fitted to the ways and slides along them.
   b. Permits transverse tool movement.
   c. Permits angular tool movement.
   d. Used to mount the cutting tool.
8. brush, your hands
9. b, c, d, and e.
10. high-speed steel (HSS)
11. carbide cutting
12. Cutting speed indicates the distance the work moves past the cutting tool, expressed in feet per minute (fpm) or meters per minute (mpm). Measuring is done on the circumference of the work.
13. Feed
14. A. 1600 rpm
    B. 200 rpm
15. 500 rpm
16. Between centers using a faceplate and dog, held in a chuck, held in a collet, and bolted to the faceplate.
17. Evaluate individually.
18. Checking centers visually by bringing their points together or by checking witness marks at base of tailstock.
19. Evaluate individually. Refer to Section 13.9.2.
20. Evaluate individually. Refer to Figure 13-76 in the text.

21. 3-jaw universal, 4-jaw independent, Jacobs, and draw-in collet. Evaluate descriptions individually. Refer to Sections 13.10.1 through 13.10.5.
22. one-third
23. To reduce chip width and prevent it from seizing (binding) in the groove.
24. Evaluate individually.

WORKBOOK ANSWERS, Pages 69–76

1. b. provides slower speeds with greater power
2. d. All of the above.
3. c. threaded spindle nose
4. d. All of the above.
5. index plate
6. lead screw
7. 2″ paintbrush
8. machine oil
9. b. to the left
10. d. It depends on the work being done.
11. e. None of the above.
12. Round nose tool
13. The irregular edge produced by grinding will crumble when used.
14. Chipbreakers
15. 685 rpm
16. 320 rpm
17. 84 rpm
18. 730 rpm
19. 186 rpm
20. a. 3-jaw universal
21. c. 4-jaw independent
22. b. Jacobs
23. collet; a separate collet is required for each different size or shape of stock
24. Using a dial indicator.
25. c. 4-jaw independent, 3-jaw universal
26. Be sure to remove the chuck key before turning on the machine.
27. c. bent-tail safety
28. b. bent-tail standard
29. a. clamp-type
30. combination drill
31. Eccentric diameters will result if the headstock center does not run true.
32. b. in either direction
33. That the cutter is slightly above center.
34. That the cutter is below center.
35. parting or cutoff
36. Long work should be center drilled and supported with a tailstock center.
37. A. Motor and gear train cover
    B. Carriage handwheel
    C. Thread and feed selector lever
    D. Quick-change gearbox
    E. Selector knob
    F. Lead screw direction lever
    G. Motor control lever
    H. Backgear handwheel
    I. Backgear control knob
    J. Headstock
    K. Variable speed control
    L. Spindle
    M. Carriage saddle
    N. Tool post
    O. Compound rest
    P. Dead center
    Q. Tailstock ram
    R. Ram lock
    S. Tailstock
    T. Tailstock lock lever
    U. Handwheel
    V. Cross-slide handwheel
    W. Rack
    X. Lead screw
    Y. Bed
    Z. Threading dial
    AA. Chip pan
    BB. Storage compartment door
    CC. Leveling screw
    DD. Tailstock pedestal
    EE. Clutch and brake handle
    FF. Half-nut lever
    GG. Power feed lever
    HH. Carriage apron
    II. Headstock pedestal
38. d. move faster or slower if the carriage is engaged to the lead screw
39. d. changes spindle speed
40. e. None of the above.
41. e. None of the above.
42. b. engages the half-nuts for threading
43. b. engages the clutch for automatic power feed
44. a. moves the entire unit right and left on the ways
45. c. automatic power cross-feed

Lathe Operation

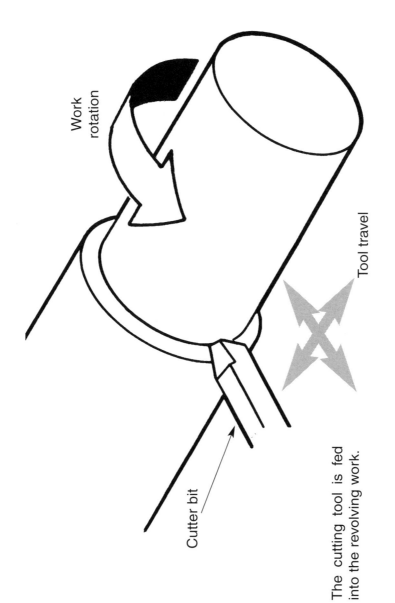

Work rotation

Tool travel

Cutter bit

The cutting tool is fed into the revolving work.

Lathe Measurement

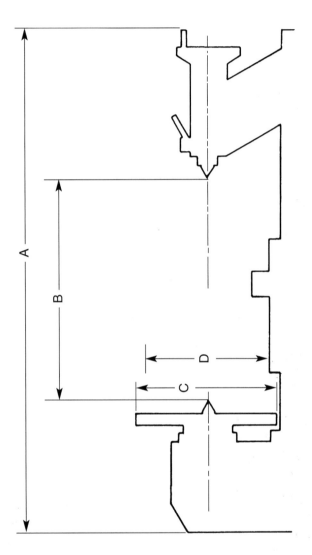

A—Length of bed. B—Distance between centers. C—Diameter of work that can be turned over the ways. D—Diameter of work that can be turned over the cross-slide.

Parts of a Lathe

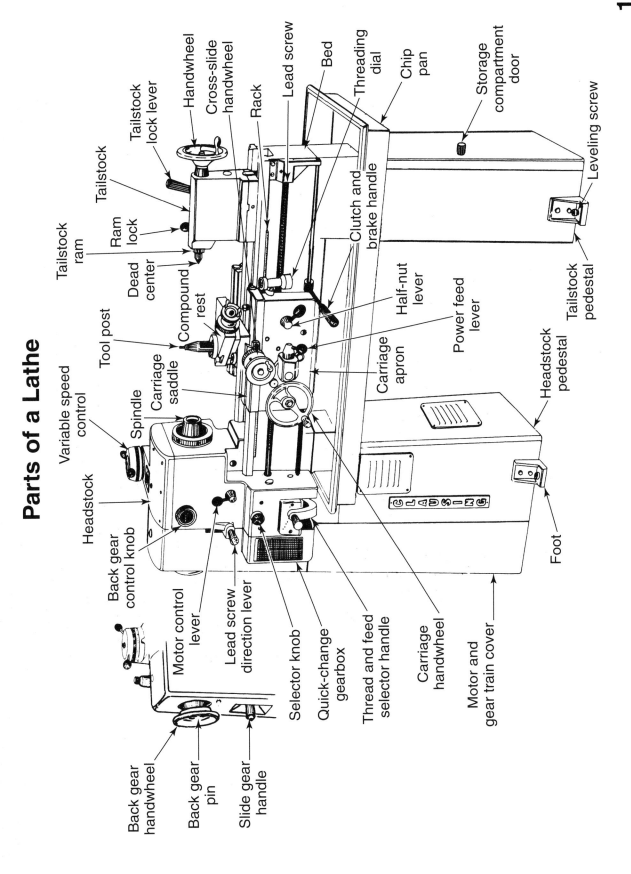

Tailstock ram

Variable speed control

Headstock

Back gear control knob

Motor control lever

Lead screw direction lever

Selector knob

Quick-change gearbox

Thread and feed selector handle

Carriage handwheel

Motor and gear train cover

Back gear handwheel

Back gear pin

Slide gear handle

Tool post

Spindle

Carriage saddle

Tailstock lock lever

Tailstock

Ram lock

Dead center

Compound rest

Handwheel

Cross-slide handwheel

Rack

Lead screw

Bed

Threading dial

Chip pan

Storage compartment door

Leveling screw

Clutch and brake handle

Half-nut lever

Carriage apron

Power feed lever

Headstock pedestal

Tailstock pedestal

Foot

High-Speed Steel Cutting Tools

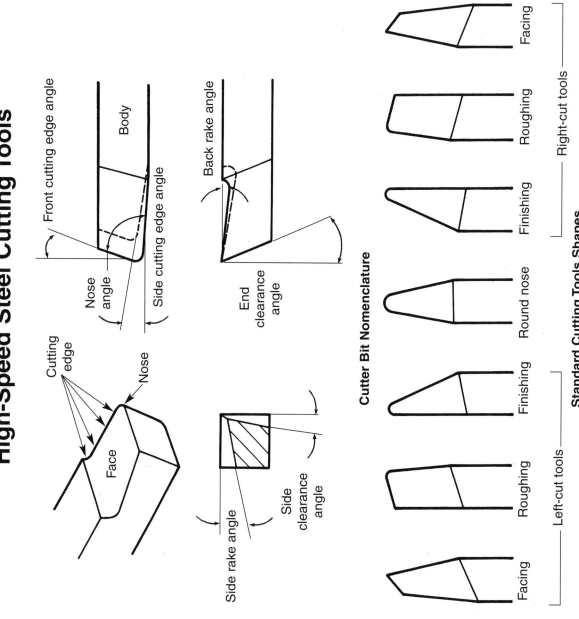

Front cutting edge angle

Body

Nose angle

Side cutting edge angle

Back rake angle

End clearance angle

Cutting edge

Nose

Face

Side rake angle

Side clearance angle

Cutter Bit Nomenclature

Facing

Roughing

Finishing

Round nose

Finishing

Roughing

Facing

Right-cut tools

Left-cut tools

Standard Cutting Tools Shapes

Sharpening HSS Cutter Bits

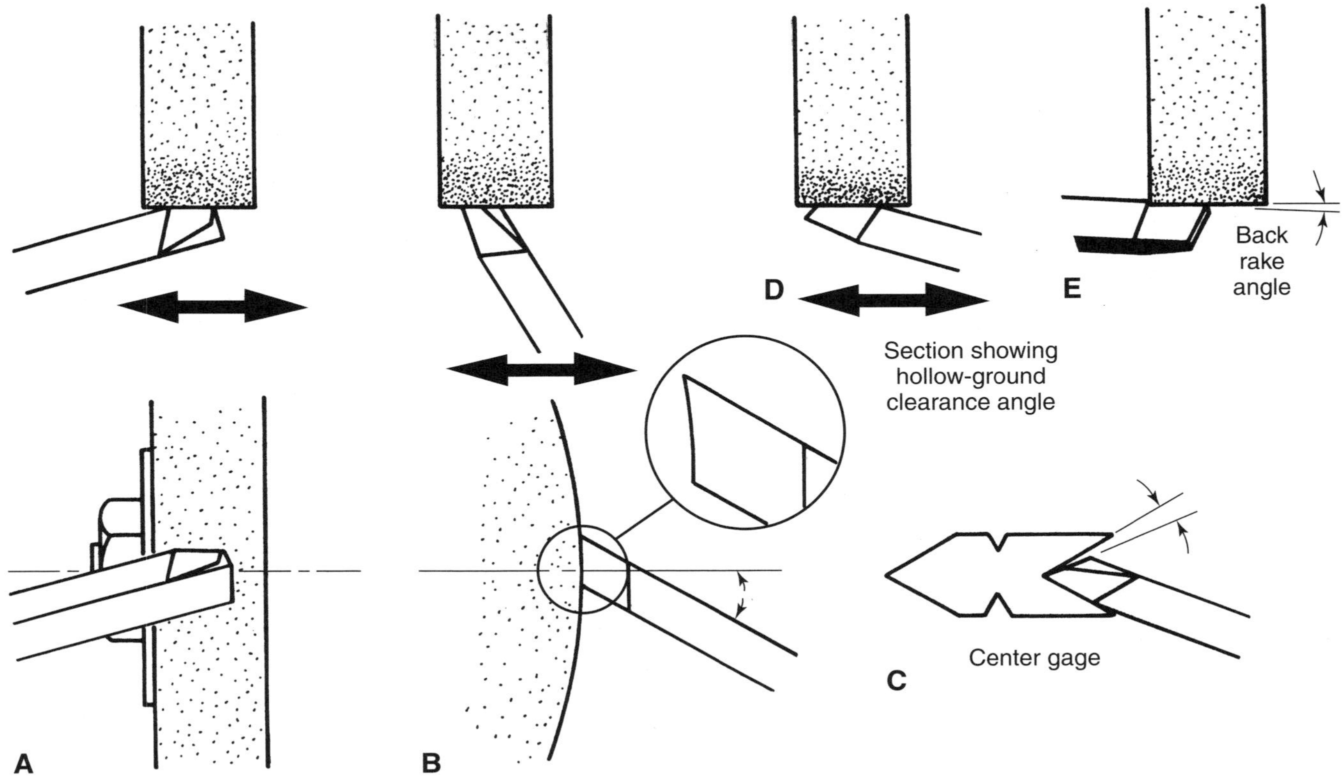
D
E
Back
rake
angle
Section showing
hollow-ground
clearance angle
Center gage
C
A
B

13-5

13-6

Using the Cutter Bit

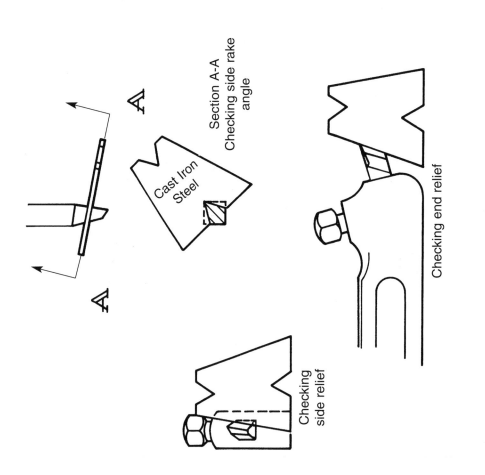

RAKE AND CLEARANCE ANGLE FOR LATHE TOOLS
(High-Speed Steel)

	Cast Iron	Low-Carbon Steel	High-Carbon Steel
Back Rake	6 – 8°	8 – 12°	4 – 6°
Side Rake	10 – 12°	14 – 18°	8 – 10°
Clearance*	6 – 9°	8 – 10°	6 – 8°
	Alloy Steels	Soft Brass	Aluminum
Back Rake	5 – 8°	0 – 2°	25 – 50°
Side Rake	10 – 15°	0 – 2°	10 – 20°
Clearance*	6 – 8°	10 – 15°	7 – 10°
	Copper		
Back Rake	10 – 12°		
Side Rake	20 – 25°		
Clearance*	6 – 8°		

* The end and side clearance angles are usually the same.

Section A-A
Checking side rake angle

Cast Iron
Steel

Checking end relief

Checking side relief

Bit gage being used to check accuracy after grinding cutter tip.

Calculating Cutting Speeds

- Cutting speeds (CS) are given in feet per minute (fpm), while the work speed is given in revolutions per minute (rpm). Thus, the peripheral speed of the work (CS) must be converted to rpm in order to determine the lathe speed required. The following formula can be used:

$$\text{rpm} = \frac{CS \times 4}{D}$$

rpm = revolutions per minute

CS = cutting speed of the particular metal being turned in feet per minute

D = diameter of the work in inches

Suggested Cutting Speeds and Feeds Using High Speed Steel (HSS) Tools				
Material to be Cut	Roughing Cut 0.01"–0.020" 0.25 mm–0.50 mm feed		Finishing Cut 0.001"–0.010" 0.025 mm–0.25 mm feed	
	fpm	mpm	fpm	mpm
Cast iron	70	20	120	36
Steel				
Low carbon	130	40	160	56
Med carbon	90	27	100	30
High carbon	50	15	65	20
Tool steel (annealed)	50	15	65	20
Brass–yellow	160	56	220	67
Bronze	90	27	100	30
Aluminum*	600	183	1000	300

The speeds for rough turning are offered as a starting point. It should be all the machine and work will withstand. The finishing feed depends upon the finish quality desired.

*The speeds for turning aluminum will vary greatly according to the alloy being machined. The softer alloys can be turned at speeds upward of 1600 fpm (488 mpm) roughing to 3500 fpm (106 mpm) finishing. High

13-7

Cutting Speed and Feed Problems

Name: _____ Date: _____ Score: _____

- Using the formula for cutting speeds, solve the following problems. Show your work in the space provided. Round your answers off to the nearest 50 rpm.

1. What spindle speed is required to finish turn 2.5″ diameter brass?

2. What spindle speed is required to finish turn 4″ diameter aluminum alloy?

3. Determine the spindle speed required to finish turn 1.25″ diameter tool steel (annealed).

4. Determine the spindle speed required to rough turn 2″ diameter cast iron.

13-8

Checking Center Alignment

13-9A

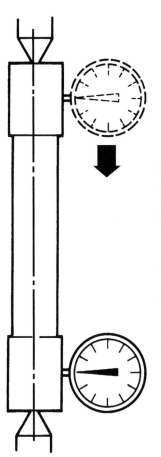

Using a Test Bar and Dial Indicator

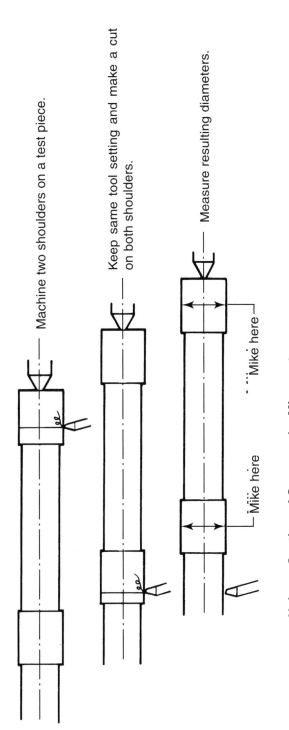

Machine two shoulders on a test piece.

Keep same tool setting and make a cut on both shoulders.

Measure resulting diameters.

Mike here

Mike here

Using a Section of Scrap and a Micrometer

Checking Center Alignment

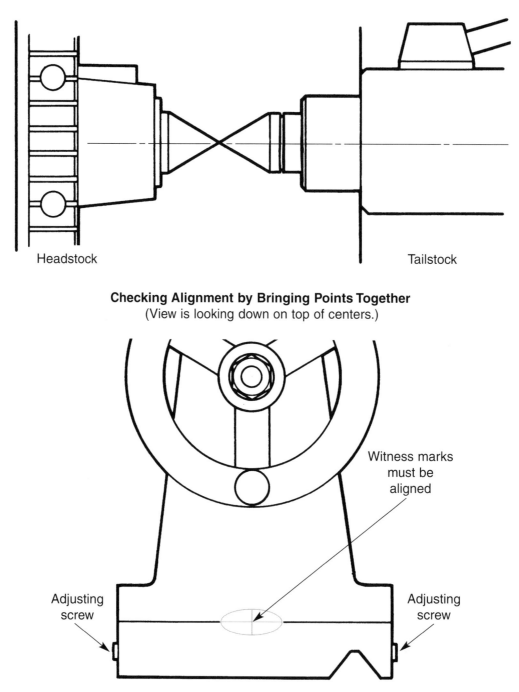

Headstock

Tailstock

Checking Alignment by Bringing Points Together
(View is looking down on top of centers.)

Witness marks
must be
aligned

Adjusting
screw

Adjusting
screw

Checking Alignment by Checking Witness Lines on Base of Tailstock

13-9B

Facing in a Chuck

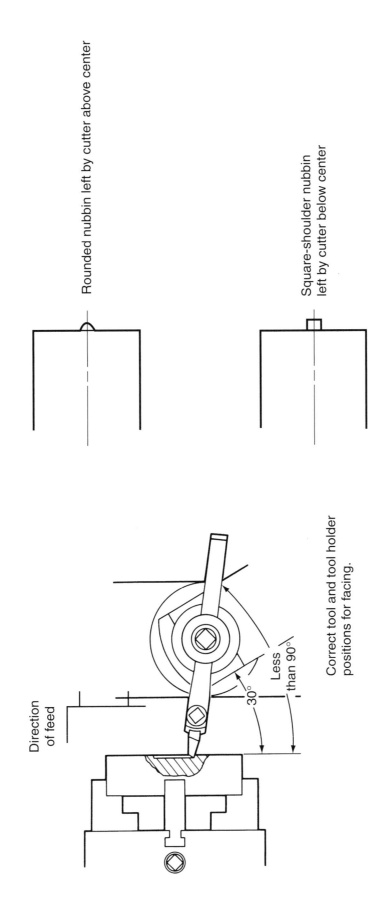

Rounded nubbin left by cutter above center

Square-shoulder nubbin left by cutter below center

Direction of feed

30°

Less than 90°

Correct tool and tool holder positions for facing.

The Lathe

Name: _____ Date: _____ Score: _____

1. The lathe operates on the principle of:
   a. The cutter revolving against the work.
   b. The cutting tool, being controllable, can be moved vertically across the work.
   c. The work rotating against the cutting tool, which is controllable.
   d. All of the above.
   e. None of the above.

   1. _____

2. The size of a lathe is determined by the _____ and the _____ of the _____.

   2. _____
   _____
   _____

3. The largest piece that can be turned between centers is equal to:
   a. The length of the bed minus the space taken up by the headstock.
   b. The length of the bed minus the space taken up by the tailstock.
   c. The length of the bed minus the space taken up by the headstock and the tailstock.
   d. All of the above.
   e. None of the above.

   3. _____

4. Into which of the following categories do the various parts of the lathe fall?
   a. Driving the lathe.
   b. Holding and rotating the work.
   c. Holding, moving, and guiding the cutting tool.
   d. All of the above.
   e. None of the above.

   4. _____

5. Explain the purpose of ways on the lathe bed. _____

   _____

   _____

   _____

   _____

   _____

6. Power is transmitted to the carriage through the feed mechanism to the quick change gearbox which regulates the amount of _____ per _____.

   6. _____
   _____

Name: _____

7. The carriage supports and controls the cutting tool. Describe each of the following parts:

a. Saddle: _____

_____

_____

b. Cross-slide: _____

_____

_____

c. Compound rest: _____

_____

_____

d. Tool post: _____

_____

_____

8. Accumulated metal chips and dirt are cleaned from the lathe with a _____, *never* with _____.

8. _____

_____

9. Which of the following actions are considered dangerous when operating a lathe?
a. Wearing eye protection.
b. Wearing loose clothing and jewelry.
c. Measuring with work rotating.
d. Operating lathe with most guards in place.
e. Using compressed air to clean machine.

9. _____

10. In most lathe operations, you will be using a single-point cutting tool made of _____.

10. _____

11. Cutting speeds can be increased 300% to 400% by using _____ tools.

11. _____

12. What does cutting speed indicate? _____

_____

_____

_____

_____

13. _____ is used to indicate the distance that the cutter moves longitudinally in one revolution of the work.

13. _____

13-11

(continued)

Name: _____

14. Calculate the cutting speeds for the following metals.   14. _____
The information furnished is sufficient to do so.

   a. Formula: rpm $\dfrac{CS \times 4}{D}$

   b. CS = Cutting speed recommended for material being machined.

   c. D = Diameter of work in inches.

Problem A: What is the spindle speed (rpm) required to finish-turn 2 1/2″ diameter aluminum alloy? A rate of 1000 fpm is the recommended speed for finish-turning the material.

Problem B: What is the spindle speed (rpm) required to rough-turn 1″ diameter tool steel? The recommended rate for rough turning the material is 50 fpm.

15. Calculating the cutting speed for metric-size material   15. _____
requires a slightly different formula.

   a. Formula: rpm = $\dfrac{CS \times 1000}{D \times 3}$

   b. CS = Cutting speed recommended for particular material being machined (steel, aluminum, etc.) in meters per minute (mpm).

   c. D = Diameter of work in millimeters (mm).

Problem: What spindle speed is required to finish-turn 200 mm diameter aluminum alloy? Recommended cutting speed for the material is 300 mpm.

Name: _____

16. Most work is machined while supported by one of four methods. List them.

_____

_____

_____

_____

17. Sketch a correctly drilled center hole.

18. A tapered piece will result, when the work is turned between centers, if the centers are not aligned. Approximate alignment can be determined by two methods. What are they?

_____

_____

19. Describe one method for checking center alignment if close tolerance work is to be done between centers. _____

_____

_____

_____

_____

20. It is often necessary to turn to a shoulder or to a point where the diameters of the work change. One of four types of shoulders will be specified. Make a sketch of each. *Make your sketches on a separate piece of paper.*
    a. Square shoulder.
    b. Angular shoulder.
    c. Filleted shoulder.
    d. Undercut shoulder.

13-11

(continued)

Name: _____

21. What are the four types of lathe chucks most commonly used? Describe the characteristics of each. _____

_____

_____

_____

_____

_____

_____

22. When using the parting tool, the spindle speed of the machine is about _____ the speed used for conventional turning.

22. _____

23. Why is a concave rake ground on top of the cutter when used for parting operations?

_____

_____

_____

24. There are many safety precautions that must be observed when operating a lathe. List what *you* consider the five most important. _____

_____

_____

_____

_____

_____

_____

_____

Chapter 14

Cutting Tapers and Screw Threads on the Lathe

LEARNING OBJECTIVES

After studying this chapter, students will be able to:
- ◯ Describe how a taper is turned on a lathe.
- ◯ Calculate tailstock setover for turning a taper.
- ◯ Safely set up and operate a lathe for taper turning.
- ◯ Describe the various forms of screw threads.
- ◯ Cut screw threads on a lathe.

INSTRUCTIONAL MATERIALS

Text: pages 241–260
  Test Your Knowledge Questions,
  pages 259–260
Workbook: pages 77–84
Instructor's Resource: pages 191–206
  Guide for Lesson Planning
  Research and Development Ideas
  Reproducible Masters:
    14-1 Angle Measurement and
      Conversion
    14-2 Tapers (*basic information*)
    14-3 Calculating Tailstock Setover
      (*taper per inch given*)
    14-4 Calculating Tailstock Setover
      (*taper per foot given*)
    14-5 Calculating Tailstock Setover
      (*all taper dimensions given*)
    14-6 Screw Thread Forms
      (*formulas included*)
    14-7 Screw Thread Lead and Pitch
    14-8 Cutting Action of Threading Tools
    14-9 Three-Wire Method of Measuring
      Threads
    14-10 Test Your Knowledge Questions
  Color Transparencies (Binder/CD only)

GUIDE FOR LESSON PLANNING

This chapter can be divided into two segments. Part I should cover cutting tapers on the lathe and Part II should cover cutting screw threads on the lathe. Copy and distribute Reproducible Masters 14-1 and 14-2.

Part I—Cutting Tapers on the Lathe

Set up lathes for demonstration purposes. Demonstrate the various ways tapers can be cut on a lathe.

Have students read and study pages 241–250. Review the assignment using Reproducible Masters 14-3, 14-4, and 14-5 as overhead transparencies and/or handouts. (Answers are located on page 193 of this Instructor's Resource.) Discuss the following:

- The advantages and disadvantages of the various methods used to cut tapers on a lathe.
- How to calculate tailstock setover.
- Methods used to setover the tailstock.
- Types of taper attachments and how to set them.
- How to measure tapers.

Emphasize the safety precautions that must be observed when cutting tapers.

Part II—Cutting Screw Threads on the Lathe

Prepare a lathe to cut threads. Explain and demonstrate procedures for cutting threads.

Have students read and study pages 250–259. Review the assignment after demonstrating how to set up a lathe and cut threads. Discuss the following:

- Major uses of the screw thread.
- Screw thread forms. Use Reproducible Masters 14-6 and 14-7.
- Review thread nomenclature. Use Reproducible Master 6-7.
- Setting up a lathe to cut 60° threads.
- Threading tool cutting action. Use Reproducible Master 14-8.
- How to use the thread dial.
- The three-wire method for measuring threads. Use Reproducible Master 14-9.
- How to cut Acme threads.
- How to cut internal threads.
- Why cutting fluid should be used.

Emphasize safety precautions to be observed when cutting threads on a lathe. Briefly review the demonstrations. Provide students with the opportunity to ask questions.

Technical Terms

Review the terms introduced in the chapter. New terms can be assigned as a quiz, homework, or extra credit. The following list is also given at the beginning of the chapter.

> *external threads*
> *internal threads*
> *major diameter*
> *minor diameter*
> *offset tailstock method*
> *pitch diameter*
> *setover*
> *taper attachment*
> *thread cutting stop*
> *three-wire method of measuring threads*

Review Questions

Assign *Test Your Knowledge* questions. Copy and distribute Reproducible Master 14-10 or have students use the questions on pages 259–260 in the text and write their answers on a separate sheet of paper.

Workbook Assignment

Assign Chapter 14 of the *Machining Fundamentals Workbook*.

Research and Development

Discuss the following topics in class or have students complete projects on their own.

1. Make a display board showing large scale models of Sharp V, square, and Acme screw threads.
2. Write a paper on how the first screw threads were made. If possible, include illustrations.
3. Demonstrate to the class the proper technique of machining screw threads. Illustrate how the tool can be repositioned after being resharpened and how to use the 3-wire method of measuring threads.

TEST YOUR KNOWLEDGE ANSWERS, Pages 259–260

1. Compound, offset tailstock, taper attachment, tool bit, and reamer. Evaluate list of advantages and disadvantages individually. Refer to Figure 14-3.
2. When it increases or decreases in diameter at a uniform rate.
3. A. 0.250″
   B. 0.563″
   C. 15.7 mm
4. Making adjustments, assembling parts, transmitting motion, applying pressure, and making measurements.
5. d. Cut on outside surface of piece.
6. f. Cut on inside surface of piece.
7. b. Largest diameter of thread.
8. a. Smallest diameter of thread.
9. e. Diameter of imaginary cylinder that would pass through threads at such points as to make width of thread and width of space at these points equal.
10. c. Distance from one point on a thread to a corresponding point on next thread.
11. g. Distance a nut will travel in one complete revolution of screw.
12. d. All of the above.
13. center gage, fish tail
14. thread dial
15. 29°
16. a. M = 0.520″

b. M = 0.270"
c. M = 0.415"
d. M = 0.509"

WORKBOOK ANSWERS, Pages 77–84

1. e. All of the above.
2. offset tailstock or tailstock setover
3. c. Both a and b.
4. micrometer dial
5. Lessens pressure on the tail center.
6. *Plain taper attachment.* Requires the cross-slide screw to be disengaged from the cross-slide feed nut. The cutting tool must be advanced by the compound rest feed screw.
   Telescopic taper attachment. It is not necessary to disengage the cross-slide feed nut.
7. Can only cut short tapers.
8. Measuring tapers by comparison plug and ring gages, etc.
   Direct measurement of tapers, gage blocks, and sine bar, etc.
9. c. thread cutting stop
10. Evaluate individually. Refer to Section 14.6.4.
11. d. All of the above.
12. 1/N (N = Number of threads per inch.)
13. b. the reverse of those used
14. sharpening the cutting tool and positioning it to cut the threads
15. start the next cut in the same direction
16. A device on the lathe that indicates when to engage the half-nuts to permit the tool to follow exactly in the original cut.
17. the half-nuts are not engaged
18. d. Both b and c.
19. b. in relation to the centerline of the taper

20. 0.025"
21. 0.044"
22. 1.50"
23. 0.876 mm
24. 175.0 mm
25. 57.14 mm
26. 0.324"
27. 0.384"
28. 0.457"
29. 0.515"
30. 0.763"
31. 0.899"
32. 1.154"
33. 0.257"
34. 0.225"
35. 0.153"
36. 0.580"
37. 0.645"
38. 0.771"
39. 1.281"
40. 1.411"

ANSWERS FOR REPRODUCIBLE MASTERS

14-3 Calculating Tailstock Setover (TPI given)

1. 0.149"
2. 0.380"
3. 0.034"
4. 0.278"
5. 0.066"

14-4 Calculating Tailstock Setover (TPF given)

1. 0.032"
2. 0.102"
3. 0.158"
4. 0.166"
5. 0.061"

14-5 Calculating Tailstock Setover (all dimensions given)

1. 0.417"
2. 0.563"
3. 0.563"
4. 0.560"
5. 0.556"

Angle Measurement and Conversion

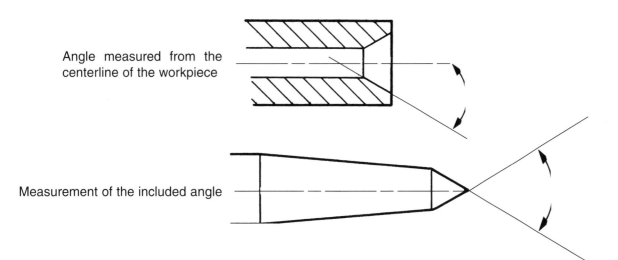

Angle measured from the centerline of the workpiece

Measurement of the included angle

Taper per Foot with Corresponding Angles		
Taper per foot	Included angle	Angle with centerline
1/16	0° 17′ 53″	0° 8′ 57″
1/8	0° 35′ 47″	0° 17′ 54″
3/16	0° 53′ 44″	0° 26′ 52″
1/4	1° 11′ 38″	0° 35′ 49″
5/16	1° 29′ 31″	0° 44′ 46″
3/8	1° 47′ 25″	0° 53′ 42″
7/16	2° 5′ 18″	1° 2′ 39″
1/2	2° 23′ 12″	1° 11′ 36″
9/16	2° 41′ 7″	1° 20′ 34″
5/8	2° 58′ 3″	1° 29′ 31″
11/16	3° 16′ 56″	1° 38′ 28″
3/4	3° 34′ 48″	1° 47′ 24″
13/16	3° 52′ 42″	1° 56′ 21″
7/8	4° 10′ 32″	2° 5′ 16″
15/16	4° 28′ 26″	2° 14′ 13″
1	4° 46′ 19″	2° 23′ 10″

Table can be used to convert taper per foot into corresponding angles for adjustment of the compound rest.

14-1

Tapers

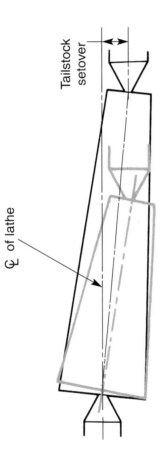

Tailstock setover

Ⅼ of lathe

Length of work causes taper to vary even though tailstock offset remains the same.

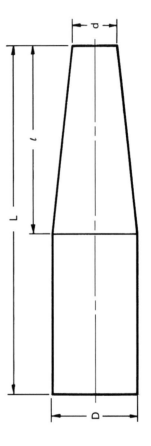

D = diameter at large end of taper; d = diameter at small end of taper; ℓ = length of taper; L = total length of piece.

Calculating Tailstock Setover

The tailstock offset must be calculated for each job because the work length plays an important role in the calculation. Information needed: TPI = Taper per inch, L = Total length of work.

Formula: When taper per inch is known, Offset = $\dfrac{L \times TPI}{2}$

1. What will be the setover for the following job? Show your work.

   TPI = 0.035″   L = 8.500″

2. What will be the setover for the following job? Show your work.

   TPI = 0.062″   L = 12.25″

3. What will be the setover for the following job? Show your work.

   TPI = 0.009″   L = 7.625″

4. What will be the setover for the following job? Show your work.

   TPI = 0.055″   L = 10.125″

5. What will be the setover for the following job? Show your work.

   TPI = 0.025″   L = 5.250″

Calculating Tailstock Setover

The tailstock offset must be calculated for each job because the work length plays an important role in the calculation. When the taper per foot (TPF) is known, it must first be converted to taper per inch (TPI). The following formula takes this into account.

Formula: When taper per foot is known, $\text{Offset} = \dfrac{L \times TPF}{24}$

1. What will be the setover for the following job? Show your work.

   TPF = 0.123″   L = 6.330″

2. What will be the setover for the following job? Show your work.

   TPF = 0.250″   L = 9.750″

3. What will be the setover for the following job? Show your work.

   TPF = 0.375″   L = 10.125″

4. What will be the setover for the following job? Show your work.

   TPF = 0.312″   L = 12.75″

5. What will be the setover for the following job? Show your work.

   TPF = 0.126″   L = 6.750″

Calculating Tailstock Setover

The tailstock setover must be calculated for each job because the work length plays an important role in the calculation. Often plans do not specify TPI, TPF, or T/mm, but do provide pertinent information. If inch dimensions are given in fractions, they must be converted to decimals.

Formula: $\text{Offset} = \dfrac{L \times (D - d)}{2 \times \ell}$

1. What will be the setover for the following job? Show your work.

   D = 2.000″   d = 1.500″ ℓ = 6.000″   L = 10.000″

2. What will be the setover for the following job? Show your work.

   D = 1.125″   d = 0.750″ ℓ = 3.000″   L = 9.000″

3. What will be the setover for the following job? Show your work.

   D = .875″   d = 0.500″ ℓ = 4.000″   L = 12.000″

4. What will be the setover for the following job? Show your work.

   D = 1.375″   d = 0.937″ ℓ = 6.000″   L = 15.00″

5. What will be the setover for the following job? Show your work.

   D = 2 1/2″   d = 1 15/16″ ℓ = 6 1/8″   L = 12 1/8″

Screw Thread Forms

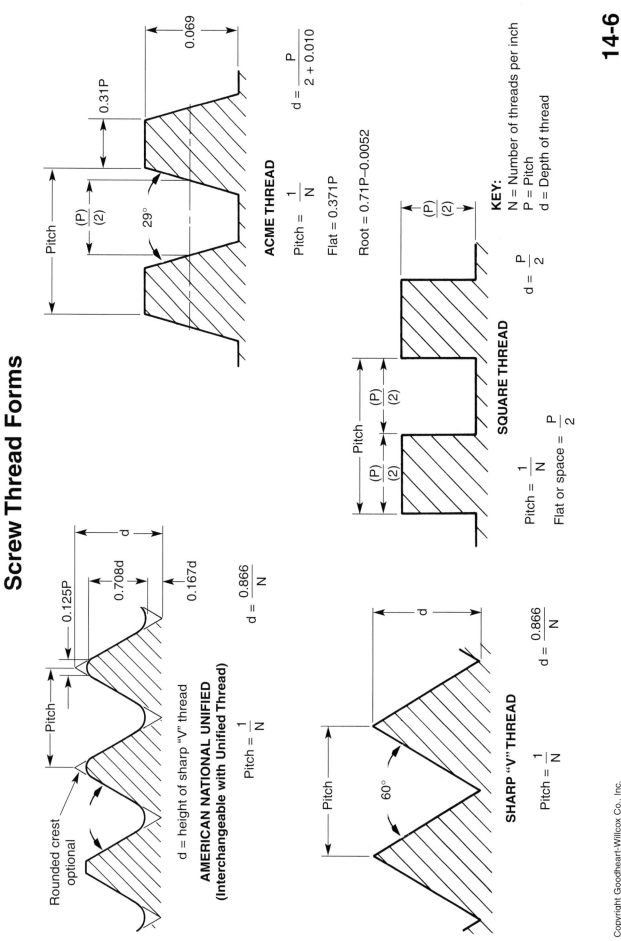

ACME THREAD

$$\text{Pitch} = \frac{1}{N}$$

$$\text{Flat} = 0.371P$$

$$\text{Root} = 0.71P - 0.0052$$

$$d = \frac{P}{2} + 0.010$$

0.069

0.31P

$\frac{(P)}{(2)}$

29°

Pitch

SQUARE THREAD

$\frac{(P)}{(2)}$

$$d = \frac{P}{2}$$

$$\text{Pitch} = \frac{1}{N}$$

$$\text{Flat or space} = \frac{P}{2}$$

Pitch

KEY:
N = Number of threads per inch
P = Pitch
d = Depth of thread

AMERICAN NATIONAL UNIFIED
(Interchangeable with Unified Thread)

$$\text{Pitch} = \frac{1}{N}$$

$$d = \frac{0.866}{N}$$

d = height of sharp "V" thread

0.125P

0.708d

0.167d

d

Pitch

Rounded crest optional

SHARP "V" THREAD

$$\text{Pitch} = \frac{1}{N}$$

$$d = \frac{0.866}{N}$$

60°

d

Pitch

Screw Thread Lead and Pitch

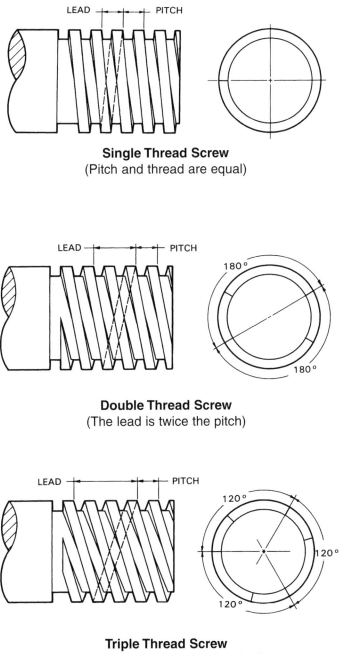

Single Thread Screw
(Pitch and thread are equal)

Double Thread Screw
(The lead is twice the pitch)

Triple Thread Screw
(The lead is three times the pitch)

14-7

14-8

Cutting Action of Threading Tools

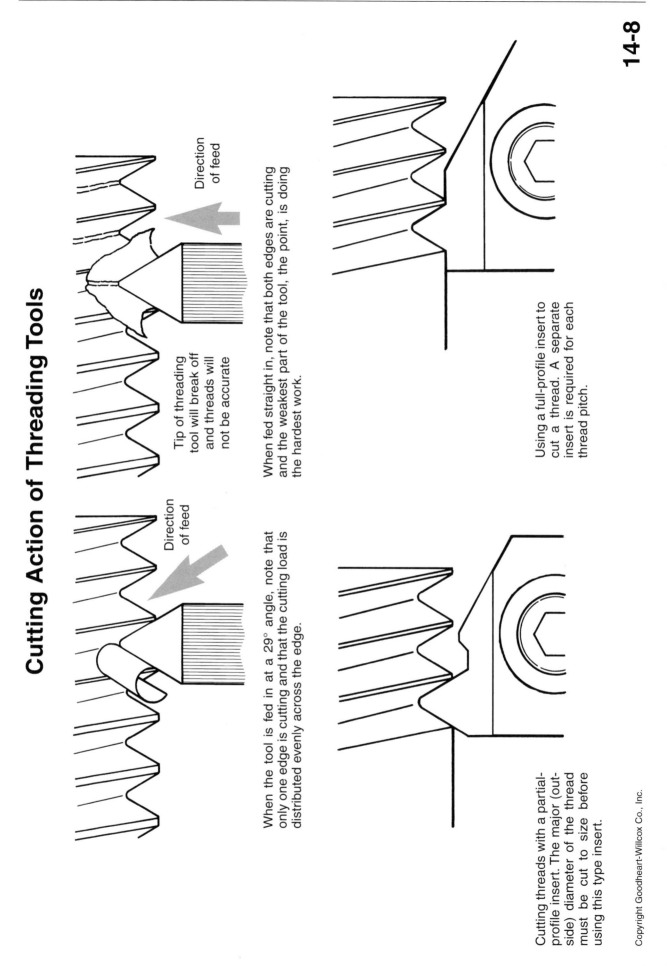

Direction of feed

Tip of threading tool will break off and threads will not be accurate

When fed straight in, note that both edges are cutting and the weakest part of the tool, the point, is doing the hardest work.

Direction of feed

When the tool is fed in at a 29° angle, note that only one edge is cutting and that the cutting load is distributed evenly across the edge.

Using a full-profile insert to cut a thread. A separate insert is required for each thread pitch.

Cutting threads with a partial-profile insert. The major (outside) diameter of the thread must be cut to size before using this type insert.

Three-Wire Method of Measuring Threads

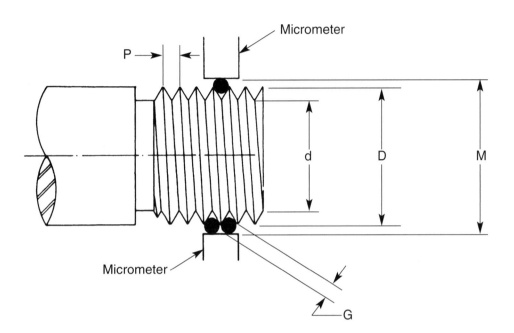

$$M = D + 3G - \frac{1.5155}{N}$$

Where: M = Measurement over the wires
D = Major diameter of thread
d = Minor diameter of thread
G = Diameter of wires
P = Pitch $= \frac{1}{N}$

N = Number of threads per inch

The smallest wire size that may be used for a given thread.

$$G = \frac{0.560}{N}$$

The largest wire size that can be used for a given thread.

$$G = \frac{0.900}{N}$$

The three-wire formula will work only if "G" is no larger or smaller than the sizes determined above. Any wire diameter between the two extremes may be used. All wires must be the same diameter.

14-9

Cutting Tapers and Screw Threads on the Lathe

Name: _____ Date: _____ Score: _____

1. There are five ways of machining tapers on a lathe. List them, with their advantages and disadvantages.

   _____

   _____

   _____

   _____

2. When is a section of material considered tapered? _____

   _____

3. Machine adjustments must be calculated for each tapering job. The information given below will enable you to calculate the necessary tailstock setover for the problems given. *Show your work in the space provided.*

Formulas: When taper per inch is known, Offset = $\dfrac{L \times TPI}{2}$

When taper per foot is known, Offset = $\dfrac{L \times TPF}{2}$

When dimensions of tapered section are known but TPI or TPF is not given,

$$\text{Offset} = \frac{L \times (D - d)}{2 \times \ell}$$

Where: TPI = Taper Per Inch   TPF = Taper Per Foot

D = Diameter at large end of taper   d = Diameter at small end of taper

ℓ = Length of taper   L = Total length of piece

Note: These formulas, except for the TPF formula, can be used when dimensions are in mm.

Problem A: What will the tailstock setover be for the following job?

Taper Per Inch = 0.125″   Total length of piece = 4.000″

Name: _____

Problem B: What will the tailstock setover be for the following job?

D = 2.50"   d = 1.75"   ℓ = 6.00"   L = 9.00"

Problem C: What will the tailstock setover be for the following job?

D = 45.0 mm   d = 25.0 mm   ℓ = 175.0 mm   L = 275.0 mm

4. Screw threads are used for many reasons. List five or more important uses.

_____

_____

_____

_____

_____

• The following questions are of the matching type. Place the letter of the correct explanation in the space provided.

_____ 5. External thread.

_____ 6. Internal thread.

_____ 7. Major diameter.

_____ 8. Minor diameter.

_____ 9. Pitch diameter.

_____10. Pitch.

_____11. Lead.

a. Smallest diameter of thread.
b. Largest diameter of thread.
c. Distance from one point on a thread to a corresponding point on next thread.
d. Cut on outside surface of piece.
e. Diameter of imaginary cylinder that would pass through threads at such points as to make width of thread and width of space at these points equal.
f. Cut on inside surface of piece.
g. Distance a nut will travel in one complete revolution of screw.

14-10

(continued)

Name: _____

12. A groove is cut at the point where a thread is to terminate.   12. _____
It is cut to the depth of the thread and serves to:
   a. provide a place to stop the threading tool after it
      makes a cut.
   b. permits a nut to be run up to the end of the thread.
   c. terminate the thread.
   d. All of the above.
   e. None of the above.

13. The tip of a cutting tool to cut a Sharp V thread is   13. _____
sharpened using a _____ to check that it is the correct
shape. This tool is frequently called a _____.   _____

14. The _____ is fitted to many lathe carriages. It meshes   14. _____
with the lead screw and is used to indicate when to
engage the half nuts to permit the thread cutting tool to
follow exactly in the original cut.

15. The compound rest is set at _____ when cutting threads   15. _____
to permit the cutting tool to shear the material better
than if it were fed straight into the work.

16. The three-wire thread measuring formula for inch-based threads is:

$$M = D + 3G - \frac{1.5155}{N}$$

Where: G = Wire diameter   D = Major diameter of thread (Convert to decimal size).

   M = Measurement over the wires   N = Number of threads per inch.

Problems: Calculate the correct measurement over the wires for the following threads. Use the wire
size given in the problem. *Show your work in the space provided.*

_____ a. 1/2-20 UNF (wire size 0.032″)

_____ b. 1/4-20 UNC (wire size 0.032″)

_____ c. 3/8-16 UNC (wire size 0.045″)

_____ d. 7/16-14 UNC (wire size 0.060″)

Chapter 15

Other Lathe Operations

LEARNING OBJECTIVES

After studying this chapter, students will be able to:
- ○ Safely set up and operate a lathe using various work-holding devices.
- ○ Properly set up steady and follower rests.
- ○ Perform drilling, boring, knurling, grinding, and milling operations on a lathe.
- ○ Demonstrate familiarity with industrial applications of the lathe.

INSTRUCTIONAL MATERIALS

Text: pages 261–280
    Test Your Knowledge Questions, page 279
Workbook: pages 85–88
Instructor's Resource: pages 207–218
    Guide for Lesson Planning
    Research and Development Ideas
    Reproducible Masters:
        15-1 Boring Tool Clearance
        15-2 Drilling on the Lathe
        15-3 Using a Drill Holder
        15-4 Knurling on the Lathe
        15-5 Mandrels
        15-6 Test Your Knowledge Questions
    Color Transparency (Binder/CD only)

GUIDE FOR LESSON PLANNING

Due to the amount of material covered, this chapter should be divided into several segments. Although it has been divided into three parts here, each classroom situation will dictate what division would work best.

Part I—Drilling, Reaming, and Boring on the Lathe

Set up lathe to demonstrate the operations. Also provide an assortment of boring bars and boring bar holders for the class to examine.

Have students/trainees read and study pages 261–264. Copy and distribute Reproducible Masters 15-1, 15-2, and 15-3. Review the assignment and demonstrate the following:

- How boring techniques differ from conventional turning.
- Selecting the proper boring bar for a job.
- Positioning a boring bar for cutting.
- Preventing chatter when using long, slender boring bars.
- Precautions to be taken when drilling, reaming, and boring.

Part II—Knurling on the Lathe

Set up a lathe to demonstrate knurling. An assortment of knurling tools should be available for examination.

Have the class read and study pages 265–267. Copy and distribute Reproducible Master 15-4. Review the reading assignment and demonstrate the following:

- Reason for knurling work.
- Different types of knurls.
- Different types of knurling tools.
- How to set up the lathe for knurling.

- Knurling difficulties and how they can be corrected.
- The use of cutting fluids when knurling.
- Precautions to be taken when knurling.

Part III—The Remainder of the Chapter

Have lathes set up to demonstrate filing and polishing, the use of steady and follower rests, using mandrels, and grinding.

Have the class read and study the remainder of the chapter. Discuss and demonstrate the following:

- Reasons filing and polishing on the lathe should be kept to a minimum.
- Safe way to file.
- When and how steady and follower rests are used.
- The use of mandrels. Use Reproducible Master 15-5.
- Grinding on the lathe.
- Precautions that must be taken when grinding on the lathe.
- Milling on the lathe.
- Special lathe attachments.
- Industrial applications of the lathe.

Briefly review the demonstrations. Provide students/trainees with the opportunity to ask questions.

Technical Terms

Review the terms introduced in the chapter. New terms can be assigned as a quiz, homework, or extra credit. The following list is also given at the beginning of the chapter.

automatic screw machine
boring
boring mills
follower rest
knurling
mandrel
reaming
steady rest
turret
turret lathe

Review Questions

Assign *Test Your Knowledge* questions. Copy and distribute Reproducible Master 15-6 or have students use the questions on page 279 in the text and write their answers on a separate sheet of paper.

Workbook Assignment

Assign Chapter 15 of the *Machining Fundamentals Workbook.*

Research and Development

Discuss the following topics in class or have students complete projects on their own.

1. Visit a local industry that uses automatic screw machines, turret lathes, and/or CNC lathes and turning centers in manufacturing their products. Prepare a short paper describing your impressions of these machines in action.
2. Borrow samples of products made on automatic screw machines and turret lathes. What characteristics, if any, do they have in common?
3. Prepare a lesson on milling on the lathe.
4. Prepare a display case or display of the various products made in your machine shop. Show several of them in various stages of construction.

TEST YOUR KNOWLEDGE ANSWERS, Page 279

1. straight, taper
2. b. An internal machining operation in which a single-point cutting tool is employed to enlarge a hole.
3. When accuracy in diameter and finish is specified.
4. knurling, gripping surface
5. It is done to remove burrs, round off sharp edges, and to blend in form cut outlines.
6. fine finish
7. When additional support is needed to prevent long and/or slender work from springing away from the cutting tool during machining. It also reduces "chattering."
8. The steady rest is bolted directly to the ways. The follower rest provides support directly in back of the cutting tool, bolts to the carriage, and follows along during the cut.
9. cat head
10. A shaft inserted through a hole in a component to support the work during machining. Used when it is necessary to machine the outside of work concentric with a hole that has been previously bored or reamed. A *solid mandrel* is made from a section of hardened steel that has been machined with a slight taper. An

expansion mandrel is machined from hardened steel and permits work with ope nings that vary from standard sizes to be turned. A *gang mandrel* is used when many pieces of the same configuration must be turned.

11. arbor press

12. tool post grinder

13. b. Cover the bed and moving parts with a heavy cloth.

WORKBOOK ANSWERS, Pages 85–88

1. Boring is employed to enlarge a hole to a specified size where a drill or reamer will not do the job.

2. reversed

3. To prevent the bottom of the tool from rubbing on the bored surface.

4. b. on center

5. tool holder

6. Any order: using a slower spindle speed; reducing tool overhang; grinding a smaller radius on the nose of the cutting tool; placing a weight on the back overhang of the boring bar; placing the tool slightly below center.

7. combination, countersink (center drill)

8. taper

9. pilot, dead center

10. Check for adequate clearance between the back of the work and the chuck face.

11. a. 0.030″ (0.8 mm)
    b. 0.010″ (0.25 mm)
    c. 0.025″ (0.6 mm)
    d. 0.015″ (0.4 mm)
    e. 0.020″ (0.5 mm)

12. One wheel is dull.

13. Move the carriage out of the way, use the left hand technique of filing, and keep the file moving.

14. centers

15. Any order: grind shafts, true lathe centers, and sharpen reamers and milling cutters.

16. Evaluate individually. Refer to Section 15.7.1.

17. The point at which the grinding wheel no longer cuts.

18. a. in opposite directions

19. vertical boring

20. turret

15-1

Boring Tool Clearance

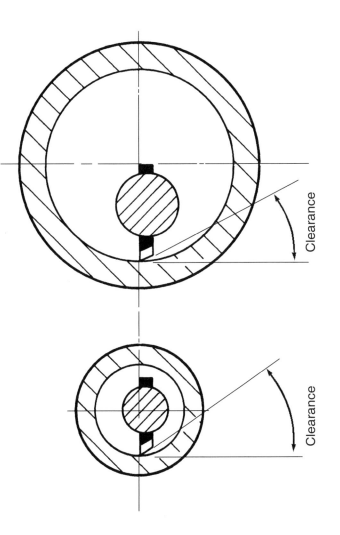

Clearance

Clearance

Tool used to bore small diameter holes requires greater front clearance to prevent rubbing.

Drilling on the Lathe

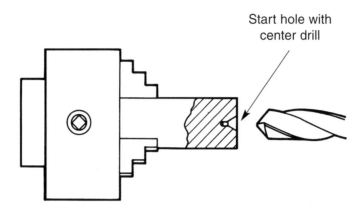

Start hole with
center drill

The drill will cut exactly on the center if the hole
is started with a center drill.

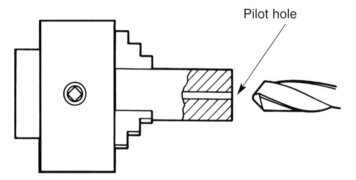

Pilot hole

Holes larger than1/2″ (12.5 mm) in diameter
require drilling of a pilot hole.

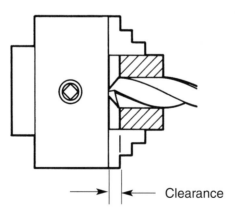

Clearance

There must be enough clearance between the
back of the work and the chuck face to permit
the drill to break through the work without damaging the chuck.

15-2

Using a Drill Holder

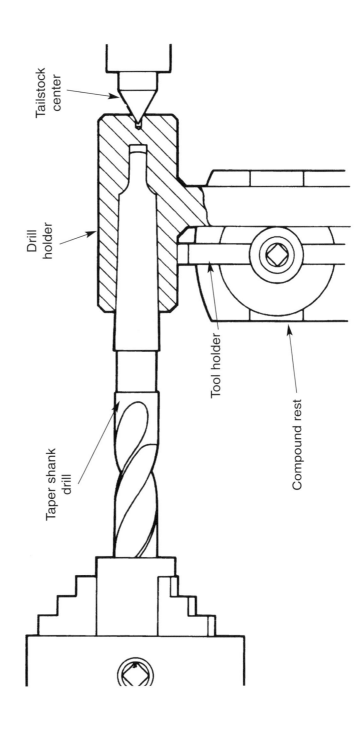

Tailstock center

Drill holder

Tool holder

Compound rest

Taper shank drill

Knurling on the Lathe

Procedure

1. Mark of section to be knurled.
2. Adjust the lathe to a slow back-geared speed and fairly rapid feed.
3. Place the knurling tool in the post. Bring it up to work. Both wheels must bear evenly on the work with their faces parallel with the centerline of the piece. See figure below.

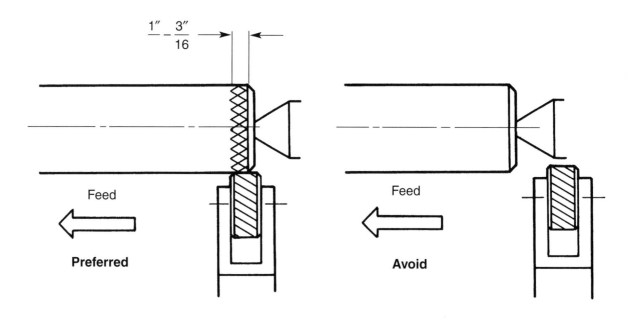

4. Start the lathe and slowly force the knurls into the work surface until a pattern begins to form. Tool travel should be *toward* the headstock whenever possible. Engage the automatic feed and let the tool travel across the work. Flood the work with cutting fluid.
5. When the knurling tool reaches the proper position, reverse spindle rotation and allow the tool to move back across the work to the starting point. Apply additional pressure to force the knurls deeper into the work.

Mandrels

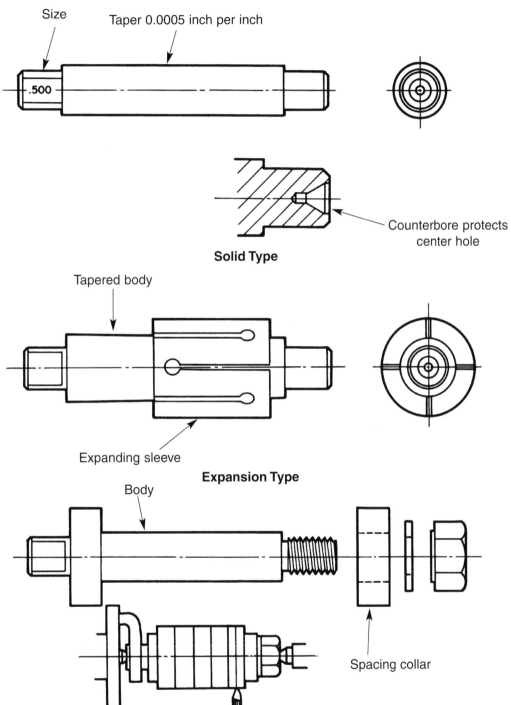

Size

Taper 0.0005 inch per inch

.500

Counterbore protects
center hole

Solid Type

Tapered body

Expanding sleeve

Expansion Type

Body

Spacing collar

Gang Mandrel

15-5

Other Lathe Operations

Name: _____ Date: _____ Score: _____

1. Drills that are used on the lathe are fitted with _____ shanks or _____ shanks.

1. _____
   _____

2. Boring is:
   a. A drilling operation.
   b. An internal machining operation in which a single-point cutting tool is employed to enlarge a hole.
   c. An external machining operation in which a single-point cutting tool is employed to reduce the diameter of a hole.
   d. All of the above.
   e. None of the above.

2. _____

3. When is reaming done?_____
   _____
   _____

4. The process of forming horizontal or diamond-shaped serrations on the circumference of the work is called _____. It is commonly done to provide a _____.

4. _____
   _____

5. When is filing on the lathe usually done? _____
   _____
   _____

6. Polishing is an operation used to produce a _____ on the work.

6. _____

7. When is a steady rest used? _____
   _____
   _____

8. What is the difference between a steady rest and a follower rest? _____
   _____
   _____

9. There are times when a shaft is unsuitable as a bearing surface and cannot be used with a steady rest. When this occurs, a _____ can be employed so the shaft can be supported with the steady rest.

9. _____

15-6
(continued)

Name: _____

10. What is a mandrel? _____

_____

_____

_____

When is it used? _____

_____

_____

_____

11. A mandrel is usually pressed into the work with an _____.

11. _____

12. Internal and external grinding can be done on a lathe with a _____.

12. _____

13. What should be done to protect the lathe from the abrasive particles that wear away from the grinding wheel?

13. _____

    a. Use a nonabrasive grinding wheel.
    b. Cover the bed and moving parts with a heavy cloth.
    c. Use a soft abrasive grinding wheel.
    d. All of the above.
    e. None of the above.

Chapter 16

Broaching Operations

LEARNING OBJECTIVES

After studying this chapter, students will be able to:
- ○ Describe the broaching operation.
- ○ Explain the advantages of broaching.
- ○ Set up and cut a keyway using a keyway broach and an arbor press.

INSTRUCTIONAL MATERIALS

Text: pages 281–284
    Test Your Knowledge Questions, page 284
Workbook: pages 89–90
Instructor's Resource: pages 219–222
    Guide for Lesson Planning
    Research and Development Ideas
    Reproducible Masters:
        16-1 How a Broaching Tool Cuts
        16-2 Test Your Knowledge Questions
    Color Transparency (Binder/CD only)

GUIDE FOR LESSON PLANNING

Have students/trainees read and study the chapter. Review the assignment using Reproducible Master 16-1 as an overhead transparency and/or handout. Discuss the following:

- The broaching process.
- Types of broaching machines.
- How a broaching tool cuts.
- Advantages of broaching.
- Demonstrate how to broach a keyway.

Technical Terms

Review the terms introduced in the chapter. New terms can be assigned as a quiz, homework, or extra credit. The following list is also given at the beginning of the chapter.

broach
broaching
burnishing
finishing teeth
keyway
pot broaching
pull broach
roughing teeth
semifinishing teeth
slab broach

Review Questions

Assign *Test Your Knowledge* questions. Copy and distribute Reproducible Master 16-2 or have students use the questions on page 284 in the text and write their answers on a separate sheet of paper.

Workbook Assignment

Assign Chapter 16 of the *Machining Fundamentals Workbook*.

Research and Development

Discuss the following topics in class or have students complete projects on their own.
1. Secure samples of work produced by broaching.
2. Research and prepare a short description of the following types of broaching machines:

a. Pot-broaching machine.

b. Continuous broaching machine.

c. Rotary broaching machine.

TEST YOUR KNOWLEDGE ANSWERS, Page 284

1. flat, round, contoured

2. It requires an opening to insert the broaching tool.

3. It is a multitoothed cutting tool. Each tooth removes only a small portion of the material being machined.

4. Any order: high productivity; can maintain close tolerances; produces good surface finishes; economical; long tool life; since equipment is automated, it can be operated by semiskilled workers.

5. burnishing (noncutting) elements

WORKBOOK ANSWERS, Pages 89–90

1. e. Both a and b.

2. Pull broach. Used for internal broaching.
   Slab broach. For external broaching.
   Pot broach. The tool is stationary and the work is moved against the tool.

3. pushed, pulled

4. A. Finishing teeth
   B. Semi-finishing teeth
   C. Roughing teeth
   D. Pilot guide

5. d. All of the above.

6. d. All of the above.

7. ram

How a Broaching Tool Cuts

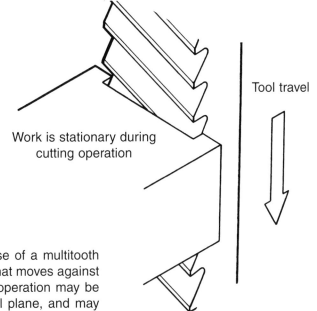

Tool travel

Work is stationary during cutting operation

Broaching involves the use of a multitooth cutting tool (the broach) that moves against the stationary work. The operation may be on a vertical or horizontal plane, and may involve making internal or external cuts.

Direction of broach travel

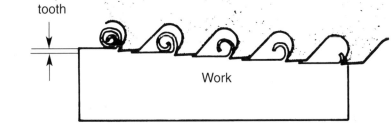

Cut per tooth

Work

Each tooth on a broaching tool removes only a small portion of the material being machined.

16-1

Broaching Operations

Name: _____ Date: _____ Score: _____

1. Broaching is a manufacturing process for machining _____ surfaces.

   1. _____

2. What does internal broaching require that external broaching does not? _____

   _____

   _____

   _____

   _____

3. What is unique about the cutting tool used on a broaching machine? _____

   _____

   _____

   _____

   _____

4. List three advantages offered by broaching. _____

   _____

   _____

   _____

   _____

   _____

5. With broaching, the machined surface can be further improved by adding _____ to the finishing end of the broach.

   5. _____

Chapter 17

The Milling Machine

LEARNING OBJECTIVES

After studying this chapter, students will be able to:
- ○ Describe how milling machines operate.
- ○ Identify the various types of milling machines.
- ○ Select the proper cutter for the job to be done.
- ○ Calculate cutting speeds and feeds.

INSTRUCTIONAL MATERIALS

Text: pages 285–316
  Test Your Knowledge Questions,
    pages 314–316
Workbook: pages 91–98
Instructor's Resource: pages 223–238
  Guide for Lesson Planning
  Research and Development Ideas
  Reproducible Masters:
    17-1 Horizontal Milling Machine
    17-2 Vertical Milling Machine
    17-3 Cutter Hand (right and left)
    17-4 Conventional and Climb Milling
    17-5 Cutting Speeds and Feeds
    17-6 Rules for Determining Speed and
      Feed
    17-7 Cutting Speed and Feed Problems
    17-8 Test Your Knowledge Questions
  Color Transparency (Binder/CD only)

GUIDE FOR PLANNING LESSON

Due to the amount of material covered, it would be advisable to divide this chapter into several segments. Although it has been divided into six parts here, each classroom situation will dictate what division would work best.

Part I—Types of Milling Machines

Set up horizontal and vertical milling machines for demonstration purposes.

Have the class read and study pages 285–293. Review the assignment using Reproducible Masters 17-1 and 17-2 as overhead transparencies and/or handouts. Discuss and demonstrate the following:

- How milling machines work.
- Types of milling machines.
- Difference between plain-type horizontal milling machine and universal-type horizontal milling machine.
- Methods of milling machine control.
- How to adjust cutting speed and feed.
- Milling operations.
- Milling safety practices.

Briefly review the demonstrations. Provide students/trainees with the opportunity to ask questions.

Part II—Milling Cutters

Set up milling machines to demonstrate facing and peripheral milling operations. Be sure the entire class is wearing approved eye protection as they view the demonstration. Also have a selection of milling cutters on hand for the class to examine. Explain how to handle milling cutters safely.

Have students/trainees read and study pages 293–304. Review the assignment, use Reproducible Master 17-3 as an overhead transparency and/or handout when discussing cutter hand. Discuss and demonstrate the following:

- Face milling and peripheral milling.
- Milling cutter classification.
- Milling cutter material.
- End mills.
- Face milling cutters.
- Fly cutters.
- Arbor milling cutters.
- Miscellaneous milling cutters.
- Care of milling cutters.
- Methods of milling.
- How to safely handle milling cutters.

Briefly review the demonstrations. Provide students/trainees with the opportunity to ask questions.

Part III—Holding and Driving Cutters

Have class read and study pages 304–308 paying particular attention to the illustrations. Review the assignment and discuss the following:

- Various types of arbors.
- Installing and removing cutter holding devices from the machines.
- Using collets.
- Care of cutter holding and driving devices.

Briefly review the demonstrations. Provide students/trainees with the opportunity to ask questions.

Part IV—Milling Cutting Speeds and Feeds

Have students/trainees read and study pages 308–310. Review the assignment emphasizing the importance of using the correct speeds and feeds. Use Reproducible Masters 17-5, 17-6, and 17-7 as overhead transparencies and/or handouts. Discuss the following:

- Calculating the correct cutting speeds and feeds.
- The purpose of cutting fluids and their importance in maintaining optimum cutting action.

Briefly review the demonstrations. Provide students/trainees with the opportunity to ask questions. Reproducible Master 17-7 contains problems that can be used as an in-class assignment of homework.

Part IV—Milling Work-Holding Attachments

An assortment of work-holding attachments should be available for class examination.

Have the class read and study pages 310–314. They should pay particular attention to the illustrations. Review the assignment and discuss the following:

- The advantages and disadvantages of the various types of vises.
- When a magnetic chuck should be used for milling operations.
- The use of the rotary and index tables.
- The dividing head and how it is set up and used.
- Safety procedures to be followed when handling heavy work-holding attachments.

Briefly review the demonstrations. Provide students/trainees with the opportunity to ask questions.

Technical Terms

Review the terms introduced in the chapter. New terms can be assigned as a quiz, homework, or extra credit. These terms are also listed at the beginning of the chapter.

> *arbor*
> *climb milling*
> *column and knee milling machine*
> *face milling*
> *horizontal spindle milling machine*
> *peripheral milling*
> *rate of feed*
> *side milling cutters*
> *traverse*
> *vertical spindle milling machine*

Review Questions

Assign *Test Your Knowledge* questions. Copy and distribute Reproducible Master 17-8 or have students use the questions on pages 314–316 and write their answers on a separate sheet of paper.

Workbook Assignment

Assign Chapter 17 of the *Machining Fundamentals Workbook.*

Research and Development

Discuss the following topics in class or have students complete projects on their own.

1. Prepare a display panel that includes samples of the various types of milling cutters. Include manufacturers' catalogs and price lists.

2. Develop a bulletin board using illustrations of the various types of milling machines. If available, include how much each machine costs.

3. The milling machine and its inventor, Eli Whitney, played an important part in developing mass production techniques. Prepare a term paper on Whitney's project of producing 10,000 muskets with interchangeable parts for the federal government in 1798. Include information on how this project led to the invention of the milling machine.

4. Make a series of posters dealing with milling machine safety.

5. Cutting fluids play an important part in any machining operation. Secure samples of cutting fluids used by industry and conduct a series of experiments to show the quality of surfaces machined dry and with the various cutting compounds. Your experiment should include milling aluminum, brass, and steel.

6. Overhaul and paint a milling machine in your training facility.

7. Demonstrate how to use a dividing head.

8. Present a video on CNC milling machines. Lead the discussion on what was seen.

9. Milling machines were the first machine tools to be automated. Do a research project on automated milling machines. Include samples of the programs, tapes, and specialized drawings used.

TEST YOUR KNOWLEDGE ANSWERS, Pages 314–316

1. fixed bed, knee and column
2. a. plain
   b. universal
   c. vertical
3. Any order: manual, semi-automatic, fully automatic, computerized. Evaluate description of each individually. Refer to Section 17.1.6.
4. measurements, adjustments
5. hand, brush
6. d. All of the above.
7. cloth, gloves
8. a. Face
   b. Peripheral
9. Solid cutter and inserted-tooth cutter.
10. Direction of rotation and helix of flutes.
11. d. Can be fed into work like a drill.
12. f. Recommended for conventional milling where plunge cutting (going into work like a drill) is *not* required.
13. c. A facing mill with a single-point cutting tool.
14. i. Mounts on a stub arbor.
15. g. Intended for machining large flat surfaces parallel to the cutter face.
16. e. Cutter with teeth located around the circumference.
17. b. Cutter with helical teeth designed to cut with a shearing action.
18. a. Has cutting teeth on the circumference and on one or both sides.
19. j. Has alternate right-hand and left-hand helical teeth.
20. h. Thin milling cutter designed for machining narrow slots and for cutoff operations.
21. fly, inserted
22. Evaluate individually. Refer to Figure 17-60.
23. Evaluate individually. Refer to Figure 17-60.
24. b. The work moves in the same direction as the rotation of the cutter.
25. a. The work is fed into the rotation of the cutter.
26. Threaded metal rod that fits through the spindle. It screws into the arbor or collet and holds it firmly in the spindle.
27. Cutting speed, feet, meters, one minute
28. Feed
29. 485 rpm, 10 ipm
30. 840 rpm, 67 ipm
31. 350 rpm
32. 190 rpm, 15 ipm
33. 380 rpm, 18 ipm
34. Any three of the following: dissipate heat; lubricate; prevent chips from sticking or fusing with the cutter teeth; flush away chips; influence the finish quality of the machined surface.

35. b. Can only be mounted parallel to or at right angles on worktable.
36. e. Has circular base graduated in degrees.
37. h. Permits compound angles (angles on two planes) to be machined without complex or multiple setups.
38. c. Needed when cutting segments of circles, circular slots, and irregular shaped slots.
39. g. Used to divide circumference of round work into equally spaced divisions.
40. f. Permits rapid positioning of circular work in 15° increments and can be locked at any angular setting.
41. d. Can only be used with ferrous metals.
42. a. Keys vise to a slot in worktable.

WORKBOOK ANSWERS,
Pages 91–98

1. e. All of the above.
2. cutter head
3. a. permits work to be positioned at several times the fastest rate indicated on the feed chart
4. d. All of the above.
5. Evaluate individually. Refer to Section 17.2.
6. Cutting speed refers to the distance, measured in feet or meters, a point (tooth) on the cutter's circumference will travel in one minute.
7. feet, meters
8. d. All of the above.
9. d. All of the above.
10. b. 3–10 times faster
11. c. Both of the above.
12. A single-point (cutting tool) face mill.
13. fed, drill
14. conventional, plunge
15. stub, end
16. c. plain milling
17. e. slab milling
18. b. side milling
19. d. staggered-tooth side
20. a. metal slitting saw

21. Feed is the rate at which the work moves into the cutter.
22. d. All of the above.
23. b. semicircular keyseats
24. d. All of the above.
25. shortest
26. Feed per tooth per revolution.
27. collars
28. Drive keys
29. Evaluate individually. Refer to section 17.7.1.
30. 250 rpm
31. 460 rpm
32. 350 rpm
33. 17800 rpm
34. 16 ipm
35. 39.6 ipm
36. is not
37. First move the workpiece clear of the cutter. Disengage the crank by withdrawing the pin from the index plate and rotating it clockwise through the section marked by the sector arms. Drop the pin into the hole at the position of the second sector arm and lock the dividing head mechanism. Next, move the sector arms in the same direction as crank rotation to catch up with the pin in the index crank. For each cut, repeat the operation.
38. d. All of the above.
39. dividing head
40. A. Vertical movement crank
    B. Saddle
    C. Longitudinal feed handwheel
    D. Swivel
    E. Overarm
    F. Motor
    G. Quill feed lever
    H. Quill feed handwheel
    I. Quill
    J. Spindle
    K. Worktable
    L. Cross traverse handwheel
    M. Base

Horizontal Milling Machine

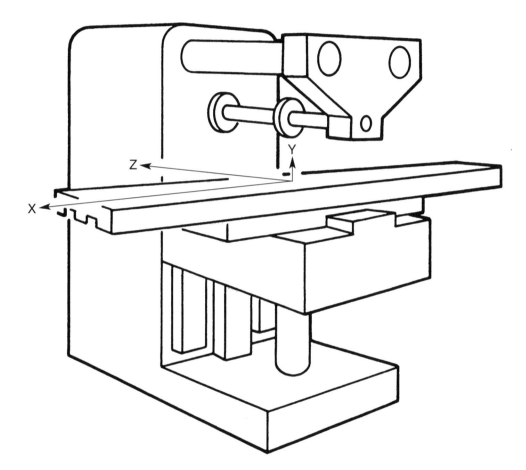

Spindle motion is assigned Z axis.

17-1

Vertical Milling Machine

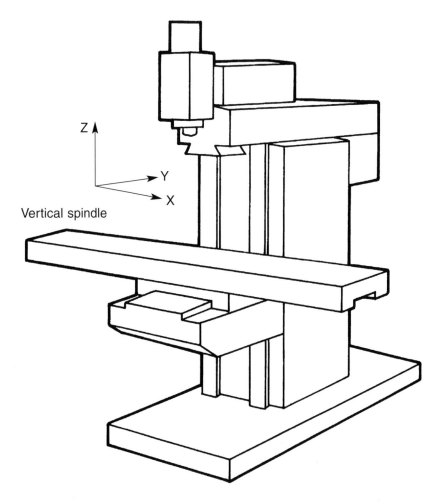

Vertical spindle

Spindle motion is assigned Z axis.

17-2

Cutter Hand

Straight shank sizes ⌀7/8 and
larger have additional flats

Driving flat ⌀3/8 and larger

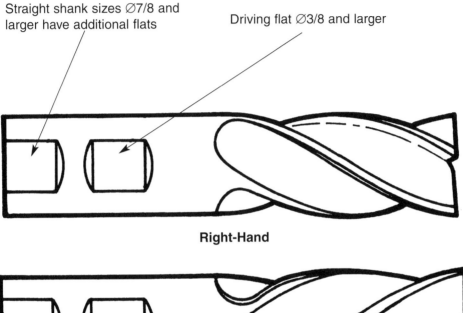

Right-Hand

Left-Hand

Cutter is right-hand if it rotates counterclockwise when viewed from cutting end. It
is left-hand if rotation is clockwise.

17-3

Conventional and Climb Milling

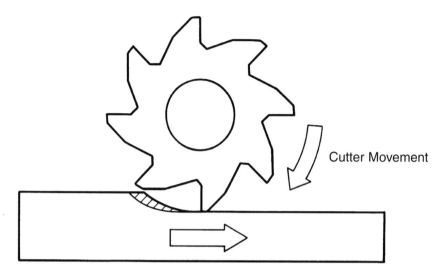

Cutter Movement

Work Movement

Conventional (up) Milling

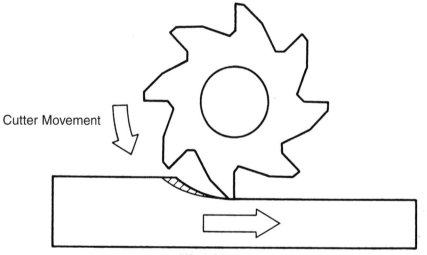

Cutter Movement

Work Movement

Climb (down) Milling

17-4

Cutting Speeds and Feeds

Material	High-speed steel cutter		Carbide cutter	
	Feet per minute	Meters per minute*	Feet per minute	Meters per minute*
Aluminum	550–1000	170–300	2200–4000	670–1200
Brass	250–650	75–200	1000–2600	300–800
Low carbon steel	100–325	30–100	400–1300	120–400
Free cutting steel	150–250	45–75	600–1000	180–300
Alloy steel	70–175	20–50	280–700	85–210
Cast iron	45–60	15–20	180–240	55–75

Reduce speeds for hard materials, abrasive materials, deep cuts, and high alloy materials. Increase speeds for soft materials, better finishes, light cuts, frail work, and setups. Start at midpoint on the range and increase or decrease speed until best results are obtained.
*Figures rounded off.

Recommended cutting speeds for milling. Speed is given in surface feet per minute (fpm) and in surface meters per minute (mpm).

Type of cutter	Material				
	Aluminum	Brass	Cast iron	Free cutting steel	Alloy steel
End mill	0.009 (0.22) 0.022 (0.55)	0.007 (0.18) 0.015 (0.38)	0.004 (0.10) 0.009 (0.22)	0.005 (0.13) 0.010 (0.25)	0.003 (0.08) 0.007 (0.18)
Face mill	0.016 (0.40) 0.040 (1.02)	0.012 (0.30) 0.030 (0.75)	0.007 (0.18) 0.018 (0.45)	0.008 (0.20) 0.020 (0.50)	0.005 (0.13) 0.012 (0.30)
Shell end mill	0.012 (0.30) 0.030 (0.75)	0.010 (0.25) 0.022 (0.55)	0.005 (0.13) 0.013 (0.33)	0.007 (0.18) 0.015 (0.38)	0.004 (0.10) 0.009 (0.22)
Slab mill	0.008 (0.20) 0.017 (0.43)	0.006 (0.15) 0.012 (0.30)	0.003 (0.08) 0.007 (0.18)	0.004 (0.10) 0.008 (0.20)	0.001 (0.03) 0.004 (0.10)
Side cutter	0.010 (0.25) 0.020 (0.50)	0.008 (0.20) 0.016 (0.40)	0.004 (0.10) 0.010 (0.25)	0.005 (0.13) 0.011 (0.28)	0.003 (0.08) 0.007 (0.18)
Saw	0.006 (0.15) 0.010 (0.25)	0.004 (0.10) 0.007 (0.18)	0.001 (0.03) 0.003 (0.08)	0.003 (0.08) 0.005 (0.13)	0.001 (0.03) 0.003 (0.08)

Increase or decrease feed until the desired surface finish is obtained.
Feeds may be increased 100 percent or more depending upon the rigity of the machine and the power available, if carbide tipped cutters are used.

Recommended feed rates in inches per tooth and millimeters (shown in parentheses) per tooth for high speed steel (HSS) milling cutters.

Rules for Determining Speed and Feed

To find	Having	Rule	Formula
Speed of cutter in feet per minute (fpm)	Diameter of cutter and revolutions per minute	Diameter of cutter (in inches) multiplied by 3.1416 (π) multiplied by revolutions per minute, divided by 12	$fpm = \dfrac{\pi D \times rpm}{12}$
Speed of cutter in meters per minute	Diameter of cutter and revolutions per minute	Diameter of cutter multiplied by 3.1416 (π) multiplied by revolutions per minute, divided by 1000	$mpm = \dfrac{D(mm) \times \pi \times rpm}{1000}$
Revolutions per minute (rpm)	Feet per minute and diameter of cutter	Feet per minute, multiplied by 12, divided by circumference of cutter (πD)	$rpm = \dfrac{fpm \times 12}{\pi D}$
Revolutions per minute (rpm)	Meters per minute and diameter of cutter in millimeters (mm)	Meters per minute multiplied by 1000, divided by the circumference of cutter (D)	$rpm = \dfrac{mpm \times 1000}{\pi D}$
Feed per revolution (FR)	Feed per minute and revolutions per minute	Feed per minute, divided by revolutions per minute	$FR = \dfrac{F}{rpm}$
Feed per tooth per revolution (ftr)	Feed per minute and number of teeth in cutter	Feed per minute (in inches or millimeters) divided by number of teeth in cutter × revolutions per minute	$ftr = \dfrac{F}{T \times rpm}$
Feed per minute (F)	Feed per tooth per revolution, number of teeth in cutter, and rpm	Feed per tooth per revolution multiplied by number of teeth in cutter, multiplied by revolutions per minute	$F = ftr \times T \times rpm$
Feed per minute (F)	Feed per revolution and revolutions per minute	Feed per revolution multiplied by revolutions per minute	$F = FR \times rpm$
Number of teeth per minute (TM)	Number of teeth in cutter and revolutions per minute	Number of teeth in cutter multiplied by revolutions per minute	$TM = T \times rpm$

rpm = Revolutions per minute
T = Teeth in cutter
D = Diameter of cutter
π = 3.1416 (pi)
frm = Speed of cutter in feet per minute

TM = Teeth per minute
F = Feed per minute
FR = Feed per revolution
ftr = Feed per tooth per revolution
mpm = Speed of cutter in meters per minute

Cutting Speed and Feed

Name: _____ Date: _____ Score: _____

Refer to the *Rules for determining cutting speeds and feeds* to calculate the cutting speed and feed for specific materials.

1. Determine the proper speed and feed for a 5″ diameter
   HSS side milling cutter with 22 teeth, milling aluminum.
   Answer the following:

   1. _____

   Recommended cutting speed for aluminum
       (midpoint on range)   =   _____

   Recommended feed per tooth
       (midpoint on range)   =   _____

   Cutter diameter   =   5″

   Number of teeth on cutter   =   22

2. Determine the proper speed and feed for a 4″ diameter
   HSS helical slab cutter with 6 teeth, milling low-carbon
   steel. Answer the following:

   2. _____

   Recommended cutting speed for low-carbon steel
       (midpoint on range)   =   _____

   Recommended feed per tooth
       (midpoint on range)   =   _____

   Cutter diameter   =   4″

   Number of teeth on cutter   =   6

The Milling Machine

Name: _____ Date: _____ Score: _____

1. Milling machines fall into two broad classifications: _____ and _____ types.

   1. _____

      _____

2. There are three basic types of milling machines.

   a. A _____ type has a horizontal spindle and the work-table has three movements.

   b. A _____ type is similar to the above machine but a fourth movement has been added to the worktable to permit it to cut helical shapes.

   c. A _____ type has the spindle perpendicular or at right angles to the worktable.

   2. a. _____

      b. _____

      c. _____

3. List the four methods of machine control. Briefly describe each of them.

   _____

   _____

   _____

   _____

4. Stop the machine before making _____ and _____.

   4. _____

      _____

5. Metal chips must never be removed with your _____. Use a _____.

   5. _____

      _____

6. Treat all small cuts and skin punctures as potential sources of infection. The following should be done:

   a. Clean them thoroughly.
   b. Apply antiseptic and cover with a bandage.
   c. Promptly report the injury to your instructor.
   d. All of the above.
   e. None of the above.

   6. _____

7. Milling cutters are sharp. Protect your hands with a _____ or _____ when handling them.

   7. _____

      _____

8. Milling operations fall into two main categories:

   a. _____ milling, in which the surface being machined is parallel with the cutter face.

   b. _____ milling, in which the surface being machined is parallel with the periphery of the cutter.

   8. a. _____

      b. _____

9. What are two general types of milling cutters? _____

   _____

Name: _____

10. What is the term *"hand"* used to describe, in reference to an end mill? _____

_____

_____

- Match each term on the left with the correct description on the right.

_____11. Two-flute end mill.

_____12. Multiflute end mill.

_____13. Fly cutter.

_____14. Shell end mill.

_____15. Face milling cutter.

_____16. Plain milling cutter.

_____17. Slab cutter.

_____18. Side milling cutter.

_____19. Staggered-tooth side cutter.

_____20. Metal slitting saw.

a. Has cutting teeth on the circumference and on one or both sides.

b. Cutter with helical teeth designed to cut with a shearing action.

c. A facing mill with a single-point cutting tool.

d. Can be fed into work like a drill.

e. Cutter with teeth located around the circumference.

f. Recommended for conventional milling where plunge cutting (going into work like a drill) is *not* required.

g. Intended for machining large flat surfaces parallel to the cutter face.

h. Thin milling cutter designed for machining narrow slots and for cutoff operations.

i. Mounts on a stub arbor.

j. Has alternate right-hand and left-hand helical teeth.

21. Flat surfaces are machined with _____ or _____ tooth milling cutters.

21. _____

_____

22. Make a sketch that illustrates climb milling. *Draw your sketch below.*

23. Make a sketch illustrating conventional milling. *Draw your sketch below.*

Name: _____

24. In climb milling: 24. _____
   a. The work is fed into the rotation of the cutter.
   b. The work moves in the same direction as the rotation of the cutter.
   c. Neither of the above.

25. In conventional milling: 25. _____
   a. The work is fed into the rotation of the cutter.
   b. The work moves in the same direction as the rotation of the cutter.
   c. Neither of the above.

26. What is a draw-in bar, and how is it used? _____

   _____

   _____

27. _____ refers to the distance, measured in _____ or _____, 27. _____
   that a point (tooth) on the circumference of a cutter _____
   moves in _____. _____
   _____
   _____

28. _____ is the rate at which the work moves into the cutter. 28. _____

- Using the formulas below, find the answers for problems 29–33. Use the space provided to show your calculations.

$$\text{rpm} = \frac{\text{fpm} \times 12}{\pi\, D}$$

$$F = \text{ftr} \times T \times \text{rpm}$$

29. Calculate machine speed (rpm) and feed (F) for a 1.5″ diameter tungsten carbide 5 tooth (T) end mill when machining cast iron. Recommended cutting speed is 190 fpm. Feed per tooth (ftr) is 0.004″.

30. Determine machine speed (rpm) and feed (F) for a 2.5″ diameter HSS shell end mill with 8 teeth (T), machining aluminum. Recommended cutting speed is 550 fpm. Feed per tooth (ftr) is 0.010″.

17-8
(continued)

Name: _____

31. Calculate machine speed (rpm) for machining aluminum with a 6″ diameter HSS side milling cutter. Recommended cutting speed is 550 fpm.

32. Determine machine speed (rpm) and feed (F) for a 4″ diameter HSS side milling cutter with 16 teeth (T) milling free cutting steel. Recommended cutting speed is 200 fpm. Feed per tooth (ftr) is 0.005″.

33. Calculate machine speed (rpm) and feed (F) for a 2.5″ diameter HSS slab milling cutter with 8 teeth (T) machining brass. Recommended cutting speed is 250 fpm. Feed per tooth (ftr) is 0.006″.

34. Cutting fluids serve several purposes. List at least three of them.

_____

_____

_____

_____

_____

• Match each term on the left with the correct description on the right.

_____ 35. Flanged vise.

_____ 36. Swivel vise.

_____ 37. Universal vise.

_____ 38. Rotary table.

_____ 39. Dividing head.

_____ 40. Indexing table.

_____ 41. Magnetic chuck.

_____ 42. Vise lug.

a. Keys vise to a slot in worktable.
b. Can only be mounted parallel to or at right angles on worktable.
c. Needed when cutting segments of circles, circular slots, and irregular shaped slots.
d. Can only be used with ferrous metals.
e. Has a circular base graduated in degrees.
f. Permits rapid positioning of circular work in 15° increments and can be locked at any angular setting.
g. Used to divide circumference of round work into equally spaced divisions.
h. Permits compound angles (angles on two planes) to be machined without complex or multiple setups.

17-8

Chapter 18

Milling Machine Operations

INSTRUCTIONAL MATERIALS

Text: pages 317–352
  Test Your Knowledge Questions, pages 351–352
Workbook: pages 99–106
Instructor's Resource: pages 239–252
  Guide for Lesson Planning
  Research and Development Ideas
  Reproducible Masters:
    18-1 Mounting End Mills
    18-2 Using the Edge Finder
    18-3 Efficiency of Small Diameter Cutter
    18-4 Straddle Milling
    18-5 Types of Gears
    18-6 Gear Nomenclature
    18-7 Bevel Gear Nomenclature
    18-8 Test Your Knowledge Questions
  Color Transparencies (Binder/CD only)

GUIDE FOR LESSON PLANNING

Due to the amount of material covered, it would be advisable to divide this chapter into several segments. Although it has been divided into five parts here, each classroom situation will dictate what division would work best.

While covering each segment of this chapter, continually emphasize the safety precautions that must be observed when operating milling machines.

Part I—Vertical Milling Machine

Have the class read and study pages 317–328, paying special attention to the illustrations. Review the assignment using Reproducible Masters 18-1 and 18-2 as overhead transparencies and/or handouts. Discuss or demonstrate the following:

- Parts of the vertical milling machine.
- Cutters for vertical milling machines.
- Vertical milling machine operation.
- Methods employed to mount cutters. Use Reproducible Master 18-1.
- Methods used to align a vise.
- Demonstrate how to align a vise with a dial indicator.
- Mounting work in a vise.
- Squaring stock with a vertical milling machine.

- Machining angular surfaces.
- Positioning a cutter to mill a keyway or slot.
- Demonstrate how to use a wiggler.
- Demonstrate how to use an edge finder. Use Reproducible Master 18-2.
- Boring on a vertical milling machine.
- Milling machine care.

Part II—Horizontal Milling Machine Operations

Have the class read and study pages 328–339 paying particular attention to the illustrations. Review the assignment using Reproducible Masters 18-3 and 18-4 as overhead transparencies and/or handouts. Discuss or demonstrate the following:

- Milling flat surfaces.
- Why the smallest diameter cutter that will do the job should be used.
- How to position a cutter using a paper strip.
- Face milling.
- Side milling.
- Straddle and gang milling.
- How to position a side cutter to mill a slot in flat stock.
- How to position a side cutter to mill a slot or keyway in round stock.
- Slitting and slotting operations.
- Drilling and boring on a horizontal milling machine.
- How to align an existing hole for boring.
- Care and cleaning of a horizontal milling machine.

Part III—Cutting a Spur Gear

A selection of gears, dividing head and gear cutters should be available for examination.

Have students read and study pages 340–346, paying particular attention to the illustrations. Review the assignment using Reproducible Masters 18-5 and 18-6 as overhead transparencies and/or handouts. Discuss or demonstrate the following:

- Gear nomenclature.
- How to determine the correct gear cutter.
- Calculating the information necessary to cut a gear.

- Preparing the dividing head to cut the required number of teeth.
- How to center the gear cutter on the work.
- Cutting the gear.
- How gear teeth measurement is checked.

Part IV—Cutting a Bevel Gear

An assortment of bevel and miter gears should be available for examination.

Copy and distribute Reproducible Master 18-7. Explain how cutting a bevel gear is a much more complex operation than cutting a spur gear. Before studying the segment on how to cut a bevel gear, allow students/trainees to examine and compare bevel and spur gears. Hold an open discussion emphasizing the differences between the two.

Have students read and study pages 346–349, paying particular attention to the illustrations. Review the assignment and discuss the complexity of cutting a bevel gear with accuracy.

Part V—Industrial Applications of Milling Machines

Using the illustrations on pages 350 and 351, point out the variety of milling machine types. Point out how the machine in Figure 18-98 is radically different.

Many companies are willing to provide educational materials in the form of press kits. Obtain as much information on new milling machines as possible. In addition to printed materials, request any available photographs, slides or videos. These materials are also available at annual trade shows.

Technical Terms

Review the terms introduced in the chapter. New terms can be assigned as a quiz, homework, or extra credit. The following list is also given at the beginning of the chapter.

addendum
bevel gear
circular pitch
dedendum
diametral pitch
gang milling
slitting
slotting
spur gear
straddle milling

Review Questions

Assign *Test Your Knowledge* questions. Copy and distribute Reproducible Master 18-8 or have students use the questions on pages 351–352 and write their answers on a separate sheet of paper.

Workbook Assignment

Assign Chapter 18 of the *Machining Fundamentals Workbook.*

Research and Development

Discuss the following topics in class or have students complete projects on their own.

1. Prepare a video or slide presentation illustrating the safety precautions that must be followed when operating a vertical milling machine.
2. Demonstrate how to use a wiggler and edge finder.
3. Demonstrate how boring is done on a vertical milling machine.
4. Demonstrate how to center a cutter on round stock for milling a keyseat. Use the paper strip technique.
5. Overhaul and paint a milling machine in your training facility.
6. Demonstrate how to cut a spur gear.
7. Demonstrate how to machine helical gears.
8. Present a video on CNC milling machines. Lead the discussion on what was seen.
9. Develop a simple project to be made by an NC or CNC milling machine. Prepare the program to do the job.

TEST YOUR KNOWLEDGE ANSWERS, Pages 351–352

1. Any order: face, end.
2. two-flute
3. two-flute
4. inserted tooth
5. vernier
6. Any order: tilting the spindle head; setting the work at the specified angle in the vise; using the protractor head of the combination set to position the work.
7. edge finder, wiggler, or centering scope
8. dial indicator
9. Evaluate individually. Refer to Section 18.3.4.

10. a. Several cutters being used at same time to machine a job.
11. To prevent it from slipping during machining.
12. smallest
13. shortest
14. Evaluate individually. Refer to Section 18.5.1.
15. 0, 45, 18
16. A spur gear is a toothed wheel that has teeth that run straight across the face and are perpendicular to the sides. The teeth are shaped so contact between the mating gears is continually maintained while in operation.
17. A rack is a flat section of metal with teeth cut into it. The combination of a spur gear and a rack converts rotary motion into linear motion.
18. Evaluate individually.

WORKBOOK ANSWERS, Pages 99–106

1. Any order: milling, drilling, boring, reaming.
2. If not, a flat surface cannot be machined.
3. With a dial indicator and a special holder.
4. By wiping the vise jaws and base clean and inspecting for burrs and ticks.
5. Thin paper strips.
6. dial indicator
7. d. All of the above.
8. against
9. Evaluate individually. Refer to Section 18.4.
10. It is more efficient because it travels less distance while doing the same amount of work as a larger cutter.
11. toward
12. b. called shell mills
13. A machining technique where a pair of cutters are used to machine both sides of the work at the same time.
14. Gang milling employs mulitiple cutters to machine several surfaces in one pass.
15. climb
16. Forces the work down against the worktable.
17. In slotting, the cut is only made partway through the work.
18. are not
19. toward
20. eight

21. To change the angular direction of power between two shafts.

22. b. narrower at the small end than at the large end

23. d. All of the above.

24. Final inspection is made by running the mating gears and checking for quietness and shape of the tooth contact.

25. D_o = 5.100"

   D = 5.000"

   h_t = 0.1078"

   t_c = 0.0785"

   a = 0.050"

   Dividing head setup = 2/5 turn per tooth

   Hole series used on index plate = 18 holes in 45 hole series in index plate

   Gear cutter to be used = No. 2

26. D_o = 5.000"

   D = 4.500"

   h_t = 0.539"

   t_c = 0.392"

   a = 0.250"

   Dividing head setup = 5/18 turn per tooth

   Hole series used in index plate = 25 holes in 90 hole series in index plate

   Gear cutter to be used = No. 6

27. D_o = 1.750"

   D = 1.500"

   h_t = 0.2696"

   t_c = 0.196"

   a = 0.125"

   Dividing head setup = 5/12 turn per tooth

   Hole series used in index plate = 15 holes in 36 hole series in index plate

   Gear cutter = No. 8

28. D_o = 3.1667"

   D = 3.0000"

   h_t = 0.1798"

   t_c = 0.1309"

   a = 0.0833"

   Dividing head setup = 1 1/9 turns per tooth

   Hole series used in index plate = 9 holes in 81 hole series in index plate

   Gear cutter to be used = No. 4

29. D_o = 6.500"

   D = 6.000"

   h_t = 0.539"

   t_c = 0.393"

   a = 0.250"

   Dividing head setup = 5/24 turn per tooth

   Hole series used in index plate = 20 holes in 96 hole series in index plate

   Gear cutter to be used = No. 5

30. D_o = 19.333"

   D = 18.667"

   h_t = 0.719"

   t_c = 0.524"

   a = 0.333"

   Dividing head setup = 5/7 turn per tooth

   Hole series used in index plate = 45 holes in 63 hole series in index plate

   Gear cutter to be used = No. 2

Mounting End Mills

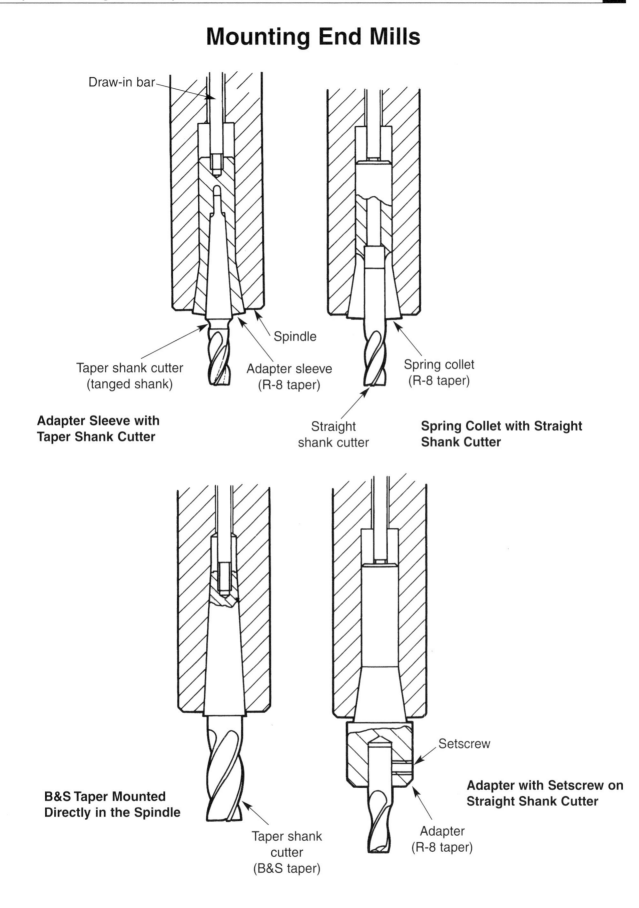

Draw-in bar

Spindle

Taper shank cutter
(tanged shank)

Adapter sleeve
(R-8 taper)

**Adapter Sleeve with
Taper Shank Cutter**

Spring collet
(R-8 taper)

Straight
shank cutter

**Spring Collet with Straight
Shank Cutter**

**B&S Taper Mounted
Directly in the Spindle**

Taper shank
cutter
(B&S taper)

Setscrew

**Adapter with Setscrew on
Straight Shank Cutter**

Adapter
(R-8 taper)

18-1

Using the Edge Finder

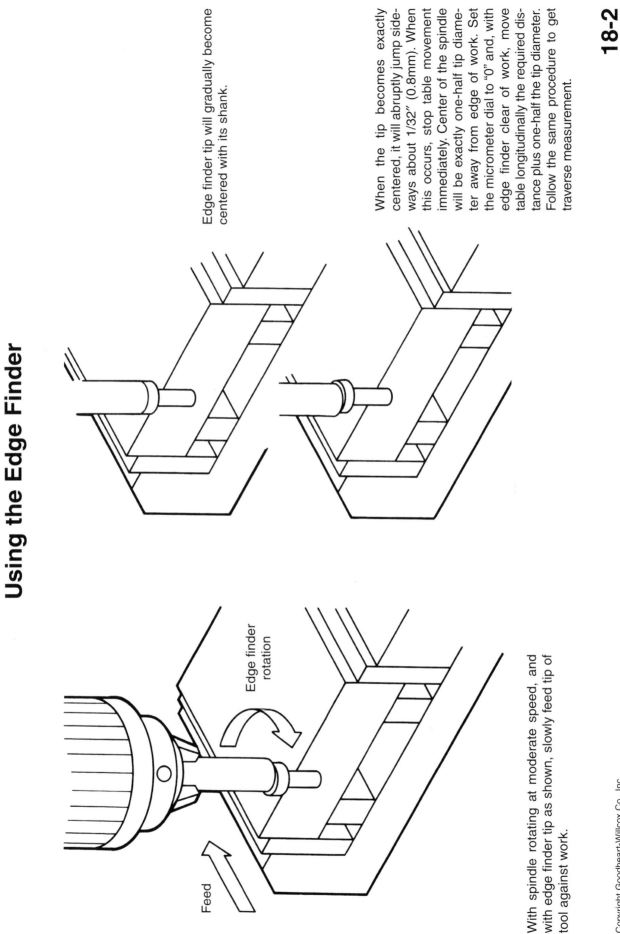

Edge finder rotation

Feed

With spindle rotating at moderate speed, and with edge finder tip as shown, slowly feed tip of tool against work.

Edge finder tip will gradually become centered with its shank.

When the tip becomes exactly centered, it will abruptly jump sideways about 1/32" (0.8mm). When this occurs, stop table movement immediately. Center of the spindle will be exactly one-half tip diameter away from edge of work. Set the micrometer dial to "0" and, with edge finder clear of work, move table longitudinally the required distance plus one-half the tip diameter. Follow the same procedure to get traverse measurement.

18-2

18-3

Efficiency of Small Diameter Cutter

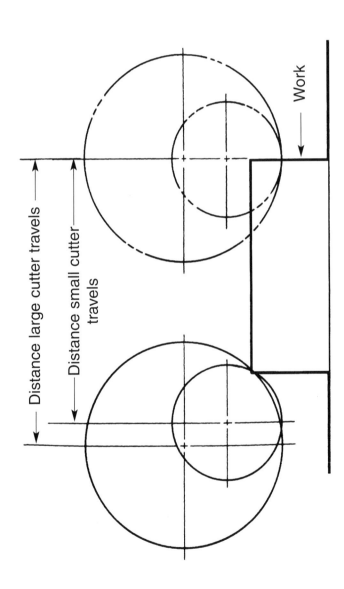

Distance large cutter travels

Distance small cutter travels

Work

Straddle Milling

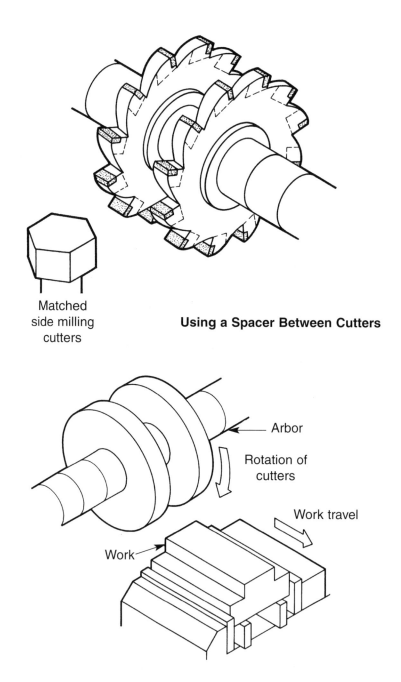

Matched
side milling
cutters

Using a Spacer Between Cutters

Arbor

Rotation of
cutters

Work travel

Work

Straddle Milling on Flatwork

18-4

Types of Gears

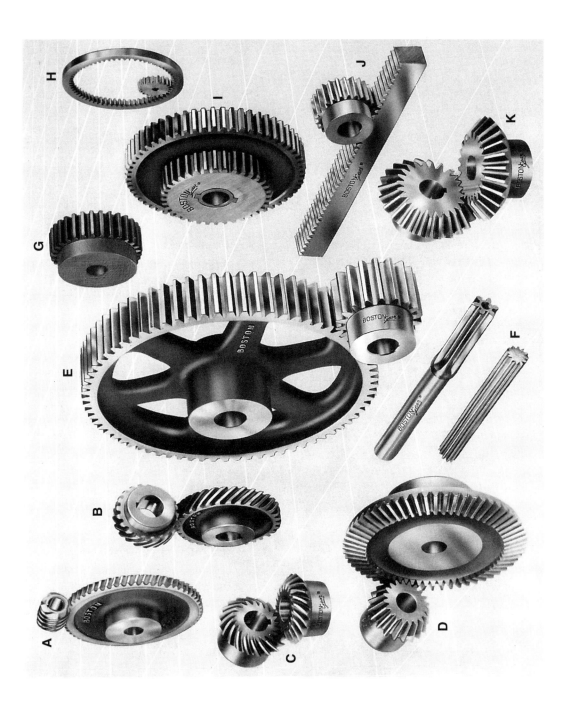

A— Worm
B— Crossed helical
C— Spiral miter
D— Bevel and pinion
E— Gear and pinion
F— Rolled pinion
G— Spur
H— Internal gear and pinion
I— Double
J— Rack and pinion
K— Miter

Gear Nomenclature

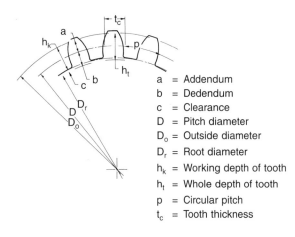

a = Addendum
b = Dedendum
c = Clearance
D = Pitch diameter
D_o = Outside diameter
D_r = Root diameter
h_k = Working depth of tooth
h_t = Whole depth of tooth
p = Circular pitch
t_c = Tooth thickness

- **Addendum** (a): The distance the tooth extends *above* the pitch circle.

$$a = \frac{1}{P} \ or \ a = \frac{D}{N} \ or \ a = \frac{D_o}{N + 2}$$

- **Circular pitch** (p): The distance, measured on the pitch circle, between similar points on adjacent teeth.

$$p = \frac{\pi p}{P} \ or \ p = \frac{\pi D}{N}$$

- **Clearance** (c): The difference between the working depth and the whole depth of a gear tooth. The amount by which the dedendum on a given gear exceeds the addendum of the mating gear.

$$c = \frac{0.157}{P}$$

- **Dedendum** (b): The distance the tooth extends *below* the pitch circle.

$$b = \frac{1.157}{P}$$

- **Distance between centers of two mating gears** (C): This distance may be calculated by adding the number of teeth of both gears and dividing one-half that sum by the diametral pitch.

$$C = \frac{N_1 + N_2}{2 \div P}$$

N_1 = Number of teeth on first gear.
N_2 = Number of teeth on second gear.

- **Diametral pitch** (P): The number of teeth per inch of pitch diameter.

$$P = \frac{N}{P} \ or \ P = \frac{N + 2}{D_o} \ or \ P = \frac{\pi}{P}$$

- **Number of teeth** (N): The number of teeth on a gear.

$$N = DP \ or \ N = D_o P - 2 \ or \ N = \frac{\pi D}{p}$$

- **Outside diameter** (D_o): Diameter or size of the gear blank.

$$D_o = D + 2_a \ or \ D_o = \frac{N}{P} + 2\left(\frac{1}{p}\right) \ or \ D_o = \frac{N + 2}{P}$$

- **Pitch circle:** An imaginary circle located approximately half the distance from the roots and tops of the gear teeth. It is tangent to the pitch circle of the mating gear.

- **Pitch diameter** (D): The diameter of the pitch circle.

$$D = \frac{N}{P} \ or \ D = 0.3183pN \ or \ D = \frac{D_o N}{N + 2}$$

- **Pressure angle** (θ): The angle of pressure between contacting teeth of mating gears. It represents the angle at which the forces from the teeth of one gear is transmitted to the mating tooth of another gear. Pressure angles of 14 1/2°, 20°, and 25° are standard. However, the 20° is replacing the older 14 1/2°.

- **Tooth thickness** (t_c): Thickness of the tooth at the pitch circle. The dimension used in measuring tooth thickness with Vernier gear tooth caliper.

$$t_c = \frac{1.5708}{P}$$

- **Whole depth of tooth** (h_t): Total depth of a tooth space, equal to the addendum (a) plus dedendum (b), or the depth to which each tooth is cut.

$$h_t = a + b \ or \ h_t = \frac{2.157}{P}$$

- **Working depth of tooth** (h_k): The sum of the addendum's of the two mating gears.

$$h_k = a_1 + a_2$$

Bevel Gear Nomenclature

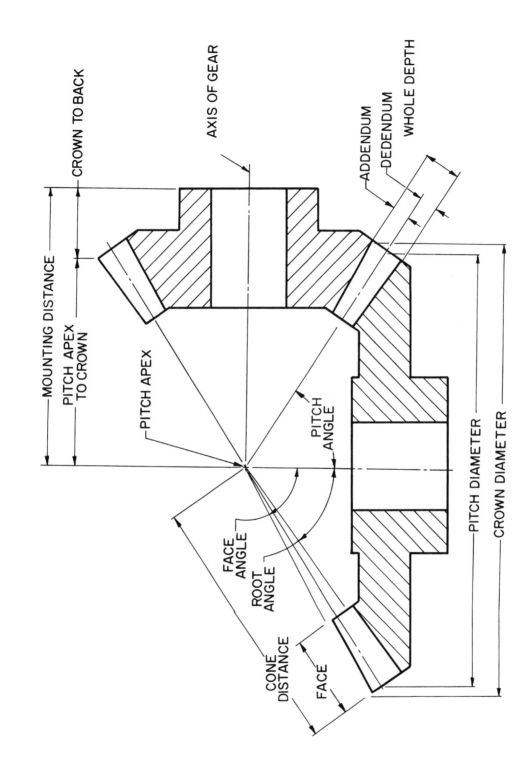

Milling Machine Operations

Name: _____ Date: _____ Score: _____

1. _____ mills and _____ mills are the cutters normally used on a vertical milling machine.

1. _____

_____

2. The _____ end mill is used when the cutter must be fed into the work like a drill.

2. _____

3. Blind holes or closed keyseats are made with a _____ end mill.

3. _____

4. Face milling cutters over 6″ (150 mm) in diameter are usually of the _____ type.

4. _____

5. A _____ scale on the spindle head of a vertical milling machine assures accurate angular settings.

5. _____

6. List three methods for machining chamfers, bevels, and tapered sections on a vertical milling machine. _____

_____

_____

_____

7. An _____ or a _____ can be used to locate the first hole of a series to be drilled on a vertical milling machine.

7. _____

_____

8. The most accurate way to align a vise on milling machine is with a _____.

8. _____

9. Explain how to center an end mill on round stock for the purpose of machining a keyseat. Use the paper strip technique. _____

_____

_____

_____

_____

10. Gang milling means:
    a. several cutters being used at same time to machine a job.
    b. two or more cutters straddling the job.
    c. several side cutters being used at same time to machine a job.
    d. All of the above.
    e. None of the above.

10. _____

Name: _____

11. Why should a milling cutter be keyed to the arbor? _____

_____

12. When sawing (slitting) thin stock, the _____ diameter cutter that provides adequate clearance should be used.

12. _____

13. In general, use the _____ arbor possible that permits adequate clearance between the arbor support and the work.

13. _____

14. Describe how to safely remove or mount a milling cutter on an arbor. _____

_____

_____

_____

_____

15. How would a dividing head be set up to cut a 100-tooth gear? The dividing head has a 40:1 ratio and the index plate has the following series of holes: 33, 37, 41, 45, 49, 53, 57.

_____Number of full turns.

_____Hole series used.

_____Number of holes in sector arm spacing.

16. What is a spur gear? _____

_____

_____

_____

17. How does a rack differ from a spur gear? _____

_____

_____

_____

18. List five precautions to be observed when operating a milling machine. _____

_____

_____

_____

_____

_____

Chapter 19

Precision Grinding

LEARNING OBJECTIVES

After studying this chapter, students will be able to:
- ○ Explain how precision grinders operate.
- ○ Identify the various-types of precision grinding machines.
- ○ Select, dress, and true grinding wheels.
- ○ Safely operate a surface grinder using various work-holding devices.
- ○ Solve common surface grinding problems.
- ○ List safety rules related to precision grinding.

INSTRUCTIONAL MATERIALS

Text: pages 353–382
   Test Your Knowledge Questions, page 381
Workbook: pages 107–112
Instructor's Resource: pages 253–268
   Guide for Lesson Planning
   Research and Development Ideas
   Reproducible Masters:
      19-1 Planer-Type Surface Grinders
      19-2 Rotary-Type Surface Grinders
      19-3 Grinding Wheel Marking System
      19-4 Grinding Wheel Shapes
      19-5 Mounting Grinding Wheels
      19-6 Creep Grinding
      19-7 Traverse Grinding
      19-8 Plunge Grinding
      19-9 Centerless Grinding
      19-10 Test Your Knowledge Questions
   Color Transparencies (Binder/CD only)

GUIDE FOR LESSON PLANNING

Due to the amount of material covered, it would be advisable to divide this chapter into several segments. Although it has been divided into seven parts here, each classroom situation will dictate what division would work best.

Part I—Types of Surface Grinders

Set up a surface grinder to demonstrate its operation. Be sure all of the class can observe and hear the demonstration. They should also be wearing approved eye protection.

Have the students/trainees read and study pages 353–356, paying particular attention to the illustrations. Review the assignment using Reproducible Masters 19-1 and 19-2 as overhead transparencies and/or handouts. Discuss the following:

- The principles of precision grinding and why it is done.
- Types of surface grinders.
- How surface grinders operate.

Part II—Work-Holding Devices

A selection of work-holding devices should be available for demonstration purposes and for the class to examine.

Have class read and study pages 356–358. Review the assignment and discuss the following:

- Types of work-holding devices used for surface grinding.

- The advantages and disadvantages of each type.
- How they operate.
- Why a demagnetizer is used.

Part III—Grinding Wheels and Cutting Fluids

Prepare a surface grinder for the class to examine. A selection of grinding wheels should be available for examination and to demonstrate how to check a grinding wheel for soundness.

Have the class read and study pages 358–363. Review the assignment using Reproducible Masters 19-3, 19-4, and 19-5 as overhead transparencies and/or handouts. Discuss the following:

- The various types and shapes of grinding wheels.
- How to determine whether a grinding wheel requires dressing.
- The grinding wheel marking system.
- The need for so many grinding wheel shapes.
- How to mount grinding wheels.
- Types of cutting fluids.
- Why cutting fluids are required for most grinding operations.
- How cutting fluids are applied.

Part IV—Grinding Applications

Prepare a surface grinder to demonstrate how to prepare the machine for operation, dress the grinding wheel, and check the machine for safe operation.

Have the class read and study pages 364–368. Review the assignment and discuss and demonstrate the following:

- Preparing a surface grinder for operation.
- The procedure for dressing a grinding wheel.
- Why a magnetic chuck is "ground-in."
- Why a piece of oiled paper is placed between the work and the magnetic chuck.
- The sequence for starting a surface grinder.
- How to use a paper strip to position the grinding wheel.
- Grinding edges square and parallel with face sides.
- Proper way to clean the surface grinder.

- Creep grinding. Use Reproducible Master 19-6.
- Grinding problems and how to correct them.
- Grinding safety.

Part V—Tool and Cutter Grinders

Prepare a tool and cutter grinder to demonstrate sharpening milling cutters.

Have the class read and study pages 368–373, paying particular attention to the illustrations. Review the assignment and discuss the following:

- Use of the tool and cutter grinder.
- Selecting the proper wheel for the sharpening operation.
- Using and adjusting tooth rest.
- Sequence for grinding plain milling cutters.
- Sequence for grinding cutters with helical teeth.
- How to grind end mills.
- How to grind form cutters.
- Sharpening taps and reamers.

Part VI—Cylindrical and Internal Grinding

Prepare a cylindrical grinder to demonstrate its operation.

Have the class read and study pages 373–376, paying particular attention to the illustrations. Review the assignment and discuss the following:

- The principle of cylindrical grinding.
- The difference between traverse and plunge grinding. Use Reproducible Masters 19-7 and 19-8.
- Holding and driving the work.
- Machine operation.
- Internal grinding operations.

Part VII—Other Grinding Operations

Have the class read and study pages 377–381, paying particular attention to the illustrations. Review the assignment and discuss the following:

- The principle of centerless grinding. Use Reproducible Master 19-9.
- The types and variations of centerless grinding.
- The principle of form grinding.

- The principle of abrasive belt machining.
- The principle of electrolytic grinding.
- Computer (CNC) grinders.

Emphasize grinding safety and the necessity of having grinder "burns" properly treated. Grinder burns are caused when the machinist's fingers or hand comes in contact with a rotating grinding wheel.

A review of the demonstrations will provide an opportunity to answer questions students/trainees may still have.

Technical Terms

Review the terms introduced in the chapter. New terms can be assigned as a quiz, homework, or extra credit. The following list is also given at the beginning of the chapter.

> centerless grinding
> creep grinding
> diamond dressing tool
> form grinding
> internal grinding
> magnetic chuck
> planer-type surface grinders
> plunge grinding
> tooth rest
> universal tool and cutter grinder

Review Questions

Assign *Test Your Knowledge* questions. Copy and distribute Reproducible Master 19-10 or have students use the questions on page 381 and write their answers on a separate sheet of paper.

Workbook Assignment

Assign Chapter 19 of the *Machining Fundamentals Workbook.*

Research and Development

Discuss the following topics in class or have students complete projects on their own.

1. Secure samples of work produced by precision grinding. Compare them with a surface roughness comparison standard and determine the degree of roughness of each sample.
2. Prepare a specimen board with surfaces finished by the various precision grinding techniques. Use illustrations to indicate the type of machine used to produce each surface.
3. Check all of the grinding wheels in the shop or lab. Discard the ones that would be dangerous to use. Design a storage rack so the good wheels can be stored safely.
4. Inspect the coolant system on the grinders in the shop or lab. Clean and make necessary repairs.
5. Prepare a list of recommendations that will improve precision grinding operations in the shop.
6. Contact grinding wheel manufacturers and request photos that show how grinding wheels are manufactured. Design a bulletin board display around the material.
7. Demonstrate how to sharpen a milling cutter.
8. Demonstrate the correct way to true and dress a grinding wheel.
9. Research the various types of coolants and the material on which they are used. Make a poster on your findings and mount it near the grinding machines.
10. Prepare a poster that lists the problems encountered with precision grinding and how they can be corrected. Mount the poster near the grinding machines.
11. Prepare a handbook on how to safely operate precision grinding machines. Duplicate it for each member in the class.

TEST YOUR KNOWLEDGE ANSWERS, Page 381

1. flat
2. Planer type and rotary type.
3. Any three of the following: magnetic chuck, universal vise, indexing head with centers, clamps, precision vise, double-faced masking tape.
4. Abrasive type, grain size, structure, grade, and bond.
5. a. Wear away as the abrasive particles become dull.
6. metallic ring
7. Loaded and glazed.
8. Cutting fluids lessen wear on the grinding wheel, help maintain accurate dimensions, affect the quality of the surface finish, and remove heat generated during the grinding operation.
9. Water-soluble chemical fluids and water-soluble-oil fluids. Polymers are also used.
10. diamond

11. glazed, loaded

12. redressing, grinding wheel

13. dirty coolant; Clean coolant system and replace coolant.

14. Conventional grinding removes the material a small amount at a time. Creep grinding does it in a single pass.

15. universal tool, cutter grinder

16. Traverse grinding and plunge grinding.

17. centerless

18. Evaluate individually. Refer to Figure 19-23.

19. abrasive belt machining

20. Electrolytic, electrochemical; Evaluate descriptions individually. Refer to Section 19.14.2.

WORKBOOK ANSWERS, Pages 107–112

1. many-tooth milling cutter as each of the abrasive particles is a separate cutting edge

2. smooth, accurate

3. c. downfeed

4. a. traverse

5. b. cross-feed

6. d. All of the above.

7. magnetic chuck

8. coolant

9. Tap it lightly with a metal rod. A solid wheel will give off a clear metallic ring.

10. Unbalanced wheels will cause irregularities on the finished ground surface.

11. Student answers will vary but may include the following: by flooding the grinding area; using a mist system; manually applying with a pressure-type oil pump can.

12. It can cause surface waviness.

13. Evaluate individually. Refer to Section 19.5.2.

14. Any of the following: clogged hydraulic lines; insufficient hydraulic fluid; hydraulic pump not functioning properly; inadequate table lubrication; cold hydraulic system; air in the system.

15. The wheel being out-of-round. It can be corrected by truing the wheel.

16. d. All of the above.

17. a. nicked or dirty chuck

18. Evaluate individually. Refer to Section 19.7.

19. Crowding

20. a. dirty coolant

21. Any or all of the following: grinding wheel may be too soft and wearing down too rapidly; tooth rest may not be mounted solidly; the arbor may not be running true on the centers.

22. workhead

23. twisting

24. radially

25. Work is mounted between centers or in a chuck and rotates while in contact with the grinding wheel.

26. b. one-third

27. d. All of the above.

28. The work is rotated against the grinding wheel. It does not have to be supported between centers. The piece is positioned on a work support blade and fed automatically between a regulating or feed wheel and a grinding wheel. The regulating wheel causes the piece to rotate and the grinding wheel does the cutting. Feed through the wheels is obtained by setting the regulating wheel at a slight angle.

29. Through feed, infeed, end feed, and internal centerless grinding.

30. shaped, contour

31. Any two of the following: removes material at a high rate; run cool and require light contact pressure; versatility; belts may be used dry or with a coolant; reduce possibility of metal distortion caused by heat; soft contact wheels and flexible belt conform to irregular shapes.

32. e. Both a and c.

33. d. All of the above.

34. d. All of the above.

35. e. None of the above.

Planer-Type Surface Grinders

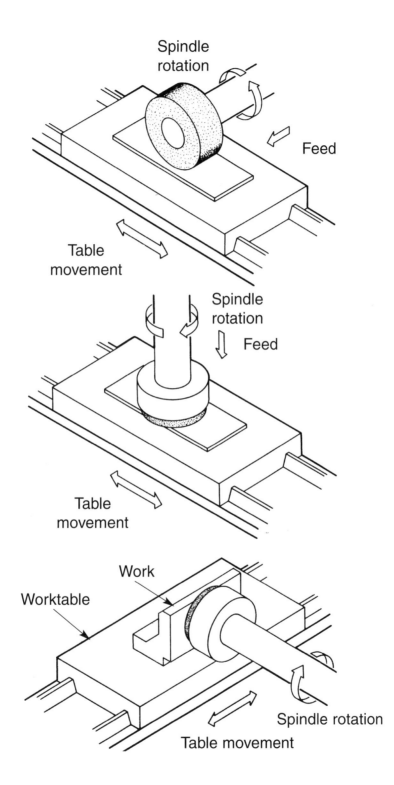

19-1

Rotary-Type Surface Grinders

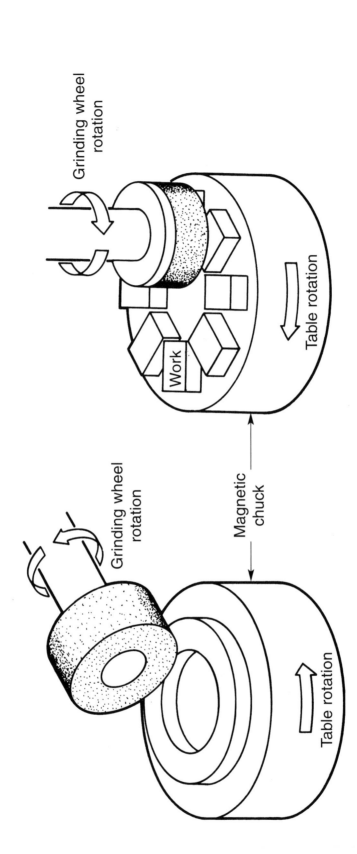

Grinding Wheel Marking System

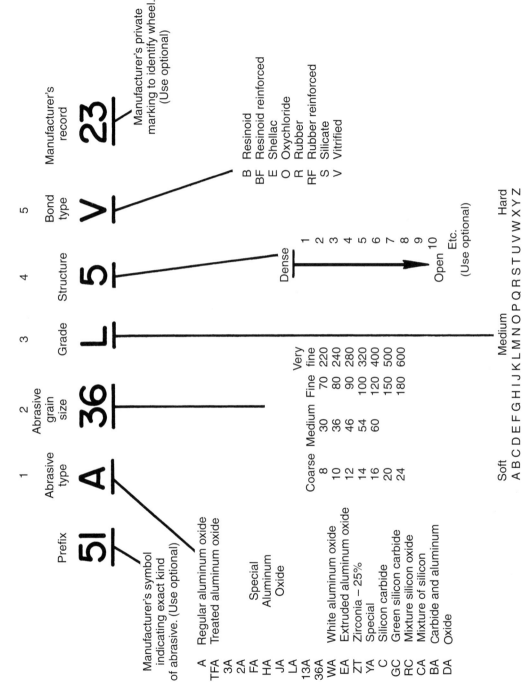

Grinding Wheel Shapes

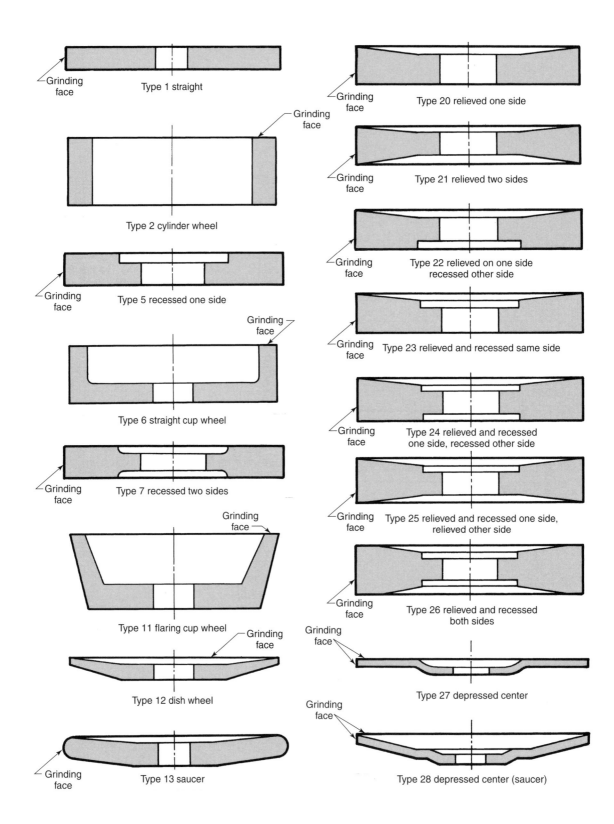

Type 1 straight

Type 2 cylinder wheel

Type 5 recessed one side

Type 6 straight cup wheel

Type 7 recessed two sides

Type 11 flaring cup wheel

Type 12 dish wheel

Type 13 saucer

Type 20 relieved one side

Type 21 relieved two sides

Type 22 relieved on one side recessed other side

Type 23 relieved and recessed same side

Type 24 relieved and recessed one side, recessed other side

Type 25 relieved and recessed one side, relieved other side

Type 26 relieved and recessed both sides

Type 27 depressed center

Type 28 depressed center (saucer)

Mounting Grinding Wheels

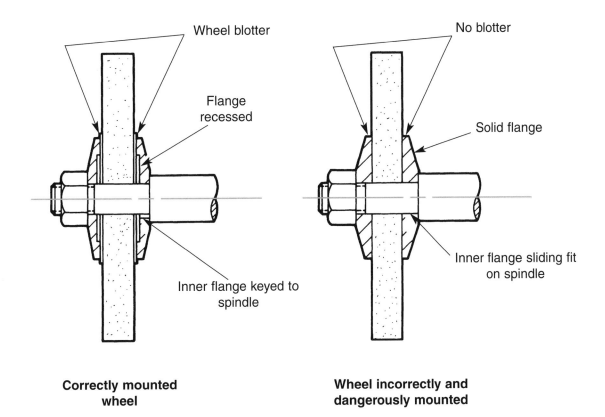

Wheel blotter

Flange recessed

Inner flange keyed to spindle

Correctly mounted wheel

No blotter

Solid flange

Inner flange sliding fit on spindle

Wheel incorrectly and dangerously mounted

19-5

Creep Grinding

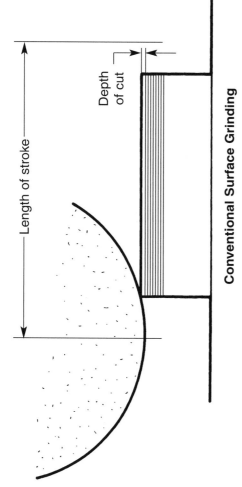

Conventional Surface Grinding

Depth of cut

Length of stroke

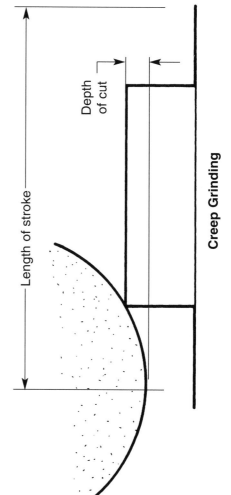

Creep Grinding

Depth of cut

Length of stroke

19-7

Traverse Grinding

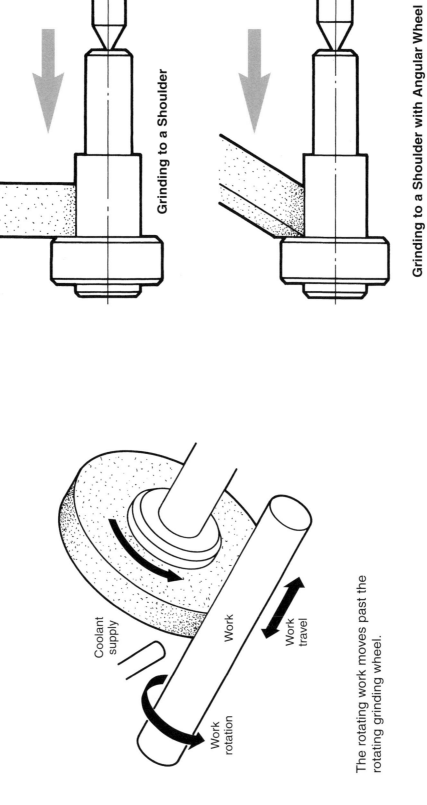

Coolant
supply

Work

Work
rotation

Work
travel

The rotating work moves past the
rotating grinding wheel.

Grinding to a Shoulder

Grinding to a Shoulder with Angular Wheel

Plunge Grinding

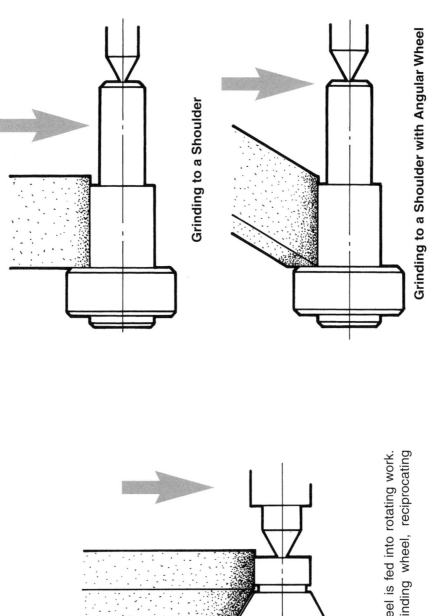

Grinding to a Shoulder

Grinding to a Shoulder with Angular Wheel

With plunge grinding, grinding wheel is fed into rotating work. Since work is no wider than grinding wheel, reciprocating motion is not needed.

19-9

Centerless Grinding

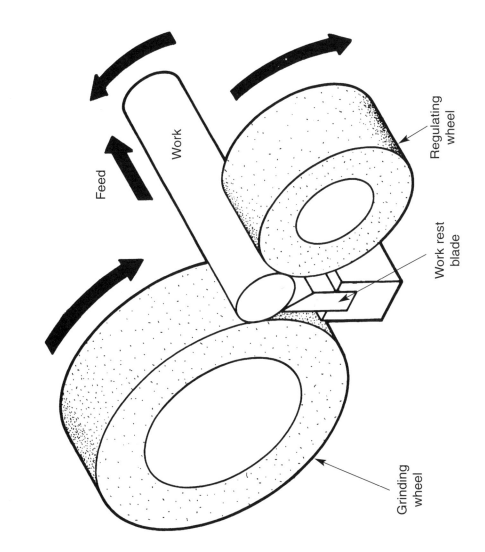

Feed

Work

Regulating wheel

Work rest blade

Grinding wheel

Precision Grinding

Name: _____ Date: _____ Score: _____

1. Industry classifies surface grinding as the grinding of   1. _____
   _____ surfaces.

2. Surface grinding operations fall into two categories. List them. _____
   _____

3. Various work-holding devices are used to hold work for surface grinding. Name three of them.
   _____

4. List five (5) factors that are distinguishing characteristics of a grinding wheel. _____
   _____
   _____
   _____

5. The ideal grinding wheel will: 5. _____
   a. wear away as the abrasive particles become dull.
   b. wear away at a predetermined rate.
   c. wear away slowly to save money.
   d. All of the above.
   e. None of the above.

6. A solid grinding wheel will give off a _____ when struck   6. _____
   lightly with a metal rod.

7. List the two conditions that commonly prevent a grinding wheel from cutting efficiently.
   _____

8. Why are cutting fluids or coolants necessary for grinding operations? _____
   _____

9. List the basic types of cutting fluids. _____

10. A _____ wheel dressing tool is usually used to true and   10. _____
    dress wheels for precision grinding.

11. Chatter and vibration marks are caused on the work   11. _____
    when the grinding wheel is _____ or _____.
                                                                  _____

12. The problems in Question 11 can be corrected by _____   12. _____
    the _____.
                                                                  _____

13. Irregular scratches on the work are usually caused by a _____ system. How can this problem be
    corrected?_____
    _____

19-10

(continued)

Name: _____

14. What is the difference between conventional grinding and creep grinding?_____

_____

_____

15. A _____ and _____ is a grinding machine designed to   15. _____
support cutters (usually milling cutters) while they are
being sharpened.   _____

16. List the two variations of cylindrical grinding. _____

_____

17. With _____ grinding, it is not necessary to support work   17. _____
between centers or mount work in a chuck while it is
being rotated against the grinding wheel.

18. Make sketches of nine standard grinding wheel shapes.

19. The grinding technique that employs a belt on which   19. _____
abrasive particles are bonded for stock removal, finishing, and polishing operations is known as _____.

20. _____ or _____ grinding is actually an electrochemical   20. _____
machining process. Describe how it is done.
                                                              _____

_____

_____

_____

_____

Chapter 20

Band Machining

LEARNING OBJECTIVES

After studying this chapter, students will be able to:
- ○ Describe how a band machine operates.
- ○ Explain the advantages of band machining.
- ○ Select the proper blade for the job to be done.
- ○ Weld a blade and mount it on a band machine.
- ○ Safely operate a band machine.

INSTRUCTIONAL MATERIALS

Text: pages 383–398
　　Test Your Knowledge Questions,
　　pages 397–398
Workbook: pages 113–116
Instructor's Resource: pages 269–278
　　Guide for Lesson Planning
　　Research and Development Ideas
　　Reproducible Masters:
　　　　20-1 Band Machining Operation
　　　　20-2 Band Machining Advantages
　　　　20-3 Blade Recommendations
　　　　20-4 Cutting Recommendations
　　　　20-5 Test Your Knowledge Questions
　　Color Transparencies (Binder/CD only)

GUIDE FOR LESSON PLANNING

Prepare the following for examination and demonstration purposes:
- Set up a band machine for operation.
- A selection of blade types (different widths, sets, and tooth forms).
- A properly welded blade.
- Welded blade sections illustrating different welding problems.
- A sampling of the type of work that can be done by band machining.

Have the class read and study the chapter. Using the various reproducible masters as overhead transparencies and/or handouts, review the assignment. Discuss and demonstrate the following:

- A description of band machining and the band machine.
- The advantages offered by band machining.
- How to select the blade to match the job.
- How to prepare and weld a blade.
- Problems that might occur when welding blades and how to correct them.
- Preparing a band machine for operation.
- Determining and setting the proper cutting speed.
- Straight sawing.
- Contour sawing.
- Making angular and internal cuts.
- Using power feed.
- Other band machining applications.
- Band machining problems and how to correct them.
- Safety precautions to be observed when band machining.

A brief review of the demonstrations will provide students/trainees the opportunity to ask questions.

Technical Terms

Review the terms introduced in the chapter. New terms can be assigned as a quiz, homework, or extra credit. The following list is also given at the beginning of the chapter.

blade guide inserts
diamond-edge band
file band
internal cuts
knife-edge blade
mist coolant
raker set
straight set
tooth form
wavy set

Review Questions

Assign *Test Your Knowledge* questions. Copy and distribute Reproducible Master 20-5 or have students use the questions on pages 397–398 and write their answers on a separate sheet of paper.

Workbook Assignment

Assign Chapter 20 of the *Machining Fundamentals Workbook*.

Research and Development

Discuss the following topics in class or have students complete projects on their own.

1. Overhaul a worn band machine in your training area. Follow procedures outlined by your instructor.
2. Collect literature on the various types of vertical band machines. Bind and catalog the collection for a technical library.
3. Secure samples of the various types of blades used on band machines. Mount and label them on a suitable display panel.
4. Visit a local industry that uses band machines. Note the type of work being done. Obtain permission to talk with machine operators to find out what their reactions are to operating a band machine. What type work is the most difficult to perform on the machine?
5. Secure a movie or video on band machining. Prepare a list of items to be discussed with the class after viewing the tape or film.
6. Demonstrate the proper way to weld a band saw blade.

7. Prepare a paper on the history of band machining.
8. Borrow samples of work produced on a band machine. Develop a display around them and explain how they were made.

TEST YOUR KNOWLEDGE ANSWERS, Pages 397–398

1. continuous saw
2. downward
3. Student answers will vary but may include three of the following: maintains sharpness; is very efficient; provides unrestricted cutting geometry; provides a built-in work-holder.
4. Blade type and blade characteristics.
5. c. Number of teeth per inch of blade or tooth spacing in millimeters.
6. a. Shape of the tooth.
7. widest
8. strong
9. annealed, brittle
10. resistant type butt
11. b. Keep it taut and tracking properly.
12. a. Flooding
    b. Mist coolant
    c. Solid lubricants
13. Straight two-dimensional cutting
14. angular
15. Threading blade through a hole drilled in piece. Weld blade and make cut.
16. band filing
17. A band machining technique that makes use of extremely high cutting speeds and heavy pressure.
18. This blade makes it possible to cut material that would tear or fray if cut with a conventional blade.
19. They are designed to cut material that is difficult or impossible to cut with a conventional toothed blade.
20. It can cut in any direction.

WORKBOOK ANSWERS, Pages 113–116

1. It uses a continuous saw blade for rapid chip removal. Each tooth is a precision cutting tool.
2. Wear leather gloves and approved eye protection.

3. c. cutting large solids

4. gage

5. b. the widest blade that will cut the desired curves

6. To assure uniform spacing after the weld is made.

7. To reduce the possibility of missing a vital point.

8. b. Roller guides

9. crown

10. By the width and pitch of the blade.

11. abrasive cloth

12. d. All of the above.

13. Any two of the following: reduce feed pressure; use finer pitch band if thin material is being cut; be sure work is held solidly as it is being fed into band; use a heavier-duty cutting fluid.

14. d. All of the above.

15. Any two of the following: remove burr on back of band where joined; if hunting back and forth against backup bearing on guide, reweld blade with back of band in true alignment; check alignment of wheels; check backup bearing, replace if worn.

16. Any of the following: use a slower cutting speed; replace the blade with a finer pitch band; be sure proper cutting fluid is used; increase feed pressure; check if band is installed with teeth pointing down.

17. Any of the following: change to a heavier band; reduce cutting speed; check wheels for damage; if blade breaks at weld, use longer annealing time; reduce heat gradually; use finer pitch blade; reduce feed pressure.

18. low

19. guides

20. A. Back
    B. Body
    C. Width
    D. Gullet depth
    E. Tooth rake angle
    F. Tooth clearance angle
    G. Tooth spacing
    H. Tooth face

Band Machining Operation

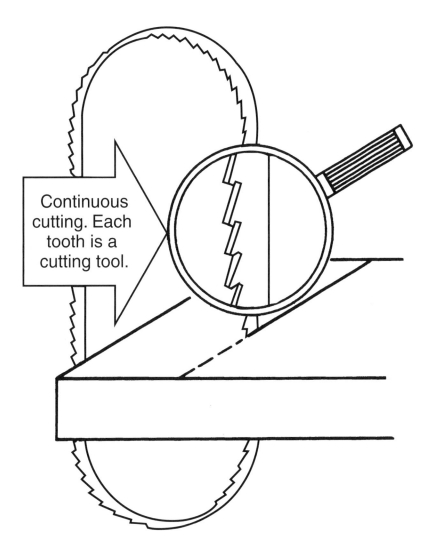

Continuous cutting. Each tooth is a cutting tool.

20-1

Band Machining Advantages

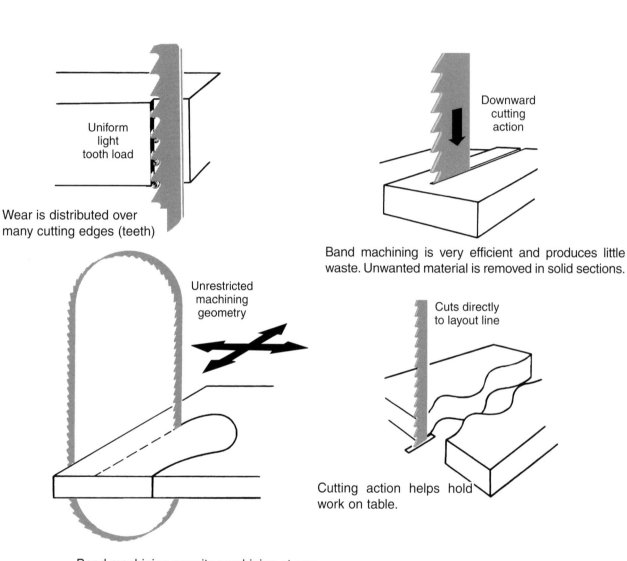

Uniform
light
tooth load

Wear is distributed over
many cutting edges (teeth)

Downward
cutting
action

Band machining is very efficient and produces little
waste. Unwanted material is removed in solid sections.

Unrestricted
machining
geometry

Cuts directly
to layout line

Cutting action helps hold
work on table.

Band machining permits machining at any
angle or direction. Cut length is almost
unlimited.

20-2

Blade Recommendations

Type of blade	Applications	Band machine
T/C Inserted tungsten carbide teeth on fatigue-resistant blade.	Heavy production and slabbing operations in tough materials.	Horizontal cutoff machines over 5 hp with positive feed. Vertical contour machines over 5 hp with positive feed.
Imperial bimetal HSS cutting edge with flex-resistant carbon-alloy back.	Mild to tough production and cutoff applications.	Horizontal cutoff machines over 1 1/2 hp with controlled feed, generally with variable-speed drives and with coolant system. Vertical contour machines over 1 1/2 hp with coolant system.
Demon M–2 HSS blade.	Heavy-duty toolroom and maintenance shop work. Full-time production applications.	Horizontal cutoff machines over 1 1/2 hp with controlled feed, generally with variable-speed drives and with coolant system. Vertical contour machines over 1 1/2 hp with coolant system.
Demon shock-resistant M–2 HSS blade specially processed for greater shock resistance.	Structurals, tubing, materials of varying cross section.	Horizontal cutoff machines over 1 1/2 hp with controlled feed, generally with variable-speed drives and with coolant system. Vertical contour machines with 1 1/2 hp with coolant system.
Dart Carbon-alloy, hard-edge, spring-tempered back blade.	Superior accuracy for light toolroom and maintenance shop applications as well as light manufacturing.	Horizontal cutoff machines under 1 1/2 hp with coolant system. Vertical contour machines under 1 1/2 hp with coolant system.
Standard carbon All-purpose, hard-edge, flexible back blade.	Light toolroom and maintenance shop applications.	Horizontal cutoff machines under 1 1/2 hp with weight feed and without coolant. Generally step speeds. Vertical contour machines under 1 1/2 hp without coolant.

Width

Blade width is from tooth tip to other edge or back.

WIDTH OF BLADE	SMALLEST RADIUS	THE WIDTH OF THE BLADE IS DETERMINED BY THE SMALLEST RADIUS TO BE CUT
1/16	1/16	
3/32	1/8	
1/8	7/32	
3/16	3/8	
1/4	5/8	
5/16	7/8	
3/8	1 1/4	
1/2	3	

20-3

Cutting Recommendations

Material thickness	Band pitch
Less than 1″ (25 mm)	10 or 14
1 to 3″ (25 to 75 mm)	6 or 8
3 to 6″ (75 to 150 mm)	4 to 6
6 to 12″ (150 to 300 mm)	2 or 3

Recommended band pitches to saw various thicknesses of material.

Material	Thickness Inches (millimeters)	Band speed Surface feet per minute (meters per minute)
Low-to-medium carbon steels	Under 1″ (25 mm)	345–360 sfm (105–110) mpm
	1″–6″ (25 mm–150 mm)	295–345 sfm (90–105) mpm
Medium-to-high carbon steels	Under 1″ (25 mm)	225–250 sfm (70–75) mpm
	1″–6″ (25 mm–150 mm)	200–225 sfm (60–70) mpm
Free machining steels	Under 1″ (25 mm)	260–395 sfm (80–120) mpm
	1″–6″ (25 mm–150 mm)	260–345 sfm (80–105) mpm
Titanium, pure and alloys	Under 1″ (25 mm)	100–115 sfm (30–35) mpm
	1″–6″ (25 mm–150 mm)	90–110 sfm (30–35) mpm

Recommended cutting speeds for selected metals and alloys.

20-4

Band Machining

Name: _____ Date: _____ Score: _____

1. Band machining makes use of a _____ blade.

1. _____

2. The cutting tool must be installed with the teeth facing _____.

2. _____

3. List three advantages band machining has over other machining techniques.

_____

_____

_____

4. What two points must be considered when selecting a blade for a specific job?

_____

5. Blade pitch refers to:

5. _____

   a. width of the blade in inches or millimeters.
   b. thickness of the blade in inches or millimeters.
   c. number of teeth per inch of blade or tooth spacing in millimeters.
   d. All of the above.
   e. None of the above.

6. Tooth form is the:

6. _____

   a. shape of the tooth.
   b. thickness of the blade.
   c. number of teeth on blade.
   d. All of the above.
   e. None of the above.

7. When making straight cuts, use the _____ blade the machine can accommodate.

7. _____

8. The joint of a properly welded blade should be as _____ as the blade itself.

8. _____

9. The blade must be _____ after welding because the joint is extremely _____ and cannot be used in this condition.

9. _____

   _____

10. A _____ welder is used to weld band machine blades.

10. _____

11. Blade tension is the pressure put on the saw band to:

11. _____

   a. cut the metal more rapidly.
   b. keep it taut and tracking properly.
   c. reduce the power needed to do the cutting.
   d. All of the above.
   e. None of the above.

20-5

(continued)

Name: _____

12. Cutting fluids are applied on the band machine in the form of:
    a. __?__ is recommended for heavy-duty sawing.
    b. __?__ is used for high-speed sawing of free machining nonferrous metals.
    c. __?__ are applied when the machine is not fitted with a built-in coolant system.

12. a. _____

   b. _____

   c. _____

13. What is the simplest form of band machining? _____

_____

_____

14. The worktable on many vertical band machines can be tilted to make _____ cuts.

14. _____

15. How can internal cuts be made on a band machine? _____

_____

_____

16. Smooth, uniformly finished surfaces are possible when the machine is fitted for _____.

16. _____

17. Describe friction sawing. _____

_____

_____

_____

18. Of what use is a knife-edge blade on a band machine? _____

_____

_____

_____

19. When are diamond-edge bands used? _____

_____

_____

_____

20. What is unique about a diamond impregnated wire band? _____

_____

_____

_____

_____

Chapter 21

Computer Numerical Control

LEARNING OBJECTIVES

After studying this chapter, students will be able to:
- ○ Define the term "numerical control."
- ○ Describe the difference between the incremental and absolute positioning methods.
- ○ Explain the operation of NC (numerical control), CNC (computer numerical control), and DNC (direct or distributed numerical control) systems.
- ○ Point out how manual and computer-aided programming is done.

INSTRUCTIONAL MATERIALS

Text: pages 399–422
  Test Your Knowledge Questions, page 421
Workbook: pages 117–122
Instructor's Resource: pages 279–292
  Guide for Lesson Planning
  Research and Development Ideas
  Reproducible Masters:
    21-1 Direct Numerical Control (DNC)
    21-2 Distributed Numerical
        Control (DNC)
    21-3 The Cartesian Coordinate System
    21-4 Axes of Machine Movements
    21-5 NC Positioning Methods
    21-6 Contour or Continuous Path
        Machining
    21-7 Mirror Image Machining
    21-8 Test Your Knowledge Questions
  Color Transparency (Binder/CD only)

GUIDE FOR LESSON PLANNING

Have the class read and study the chapter. Review the assignment using the reproducible masters as overhead transparencies and/or handouts. Discuss the following:

- The meaning of *Computer-Aided Machining Technology.*
- Numerical Control (NC) and Computer Numerical Control (CNC).
- The difference between *Direct* and *Distributed Numerical Control.* Use Reproducible Masters 21-1 and 21-2.
- The Cartesian Coordinate System. Use Reproducible Master 21-3.
- NC tool positioning methods (*absolute* and *incremental*). Use Reproducible Masters 21-4 and 21-5.
- NC movement systems, including *point-to-point, straight-cut,* and *contour or continuous path.* Use Reproducible Master 21-6.
- Mirror imaging. Use Reproducible Master 21-7.
- Programming NC machines, both manual and computer-aided.
- Computer languages.
- Adaptive control.
- Advantages and disadvantages of NC.

- Other NC applications.
- Setting up and programming the NC machine in the shop/lab.
- Demonstrating the NC machine in the shop/lab.

A brief review of the demonstrations will provide students/trainees the opportunity to ask questions.

Technical Terms

Review the terms introduced in the chapter. New terms can be assigned as a quiz, homework, or extra credit. The following list is also given at the beginning of the chapter.

absolute positioning
Cartesian Coordinate System
circular interpolation
closed loop system
continuous path system
incremental positioning
machine control unit (MCU)
open loop system
point-to-point system
straight-cut system

Review Questions

Assign *Test Your Knowledge* questions. Copy and distribute Reproducible Master 21-8 or have students use the questions on page 421 and write their answers on a separate sheet of paper.

Workbook Assignment

Assign Chapter 21 of the *Machining Fundamentals Workbook.*

Research and Development

Discuss the following topics in class or have students complete projects on their own.

1. Design and construct a simple machine that will illustrate how *numerical control* works.
2. Review up-to-date technical magazines that have articles on subjects such as NC, CNC, and robotics. Prepare a brief outline of at least one article for class discussion.
3. Visit a plant that uses automated equipment. If such a visit is not possible, show a video or film that illustrates automation.
4. If your shop/lab is fortunate enough to have an NC or CNC machine tool, ask your instructor to assign a programming problem. Prepare the program, edit and proof it,

and follow through to the finished machined part.

TEST YOUR KNOWLEDGE ANSWERS, Page 421

1. Evaluate individually. Refer to Section 21.1.
2. Evaluate individually. Refer to Section 21.2.1 and Figure 21-6.
3. Evaluate individually. Refer to Section 21.2.2 and Figures 21-10 and 21-11.
4. A sequence of instructions that tells the machine what operations to perform, and where on the material they are to be done.
5. Point-to-point, straight-cut, and contour or continuous path.
6. Evaluate individually. Refer to Figures 21-13, 21-14, and 21-15.
7. Contour or continuous path system. Geometrical complexity makes a computer mandatory when preparing programs to machine two- and three-dimensional shapes.
8. a. Machine Control Unit.
   b. A series of letters, numbers, punctuation marks, and special characters used to instruct the machine what operations to perform and where to perform them.
   c. A script containing lines of information blocks.
9. By producing a prototype made of plastic or wax. Refer to Figure 21-41.
10. Evaluate individually. Refer to Section 21.6.

WORKBOOK ANSWERS, Pages 117–122

1. Manual machining is done by the machinist moving one or more of the machine's lead and feed screws and guiding it through the various machining operations.
2. a. is the operation of the machine tool by a series of coded instructions
3. d. All of the above.
4. c. Both a and b.
5. d. All of the above.
6. The closed loop system uses an electronic feed-back device, called a transducer, to continually monitor tool position. The open loop system has no feedback for monitoring for comparing purposes. The system relies on the integrity of the control unit for accuracy.

7. The basis for NC programming. It provides a way to define movement.

8. c. Z axis

9. CNC (computer numerical control)

10. c. a fixed point of origin, or zero point

11. b. the prior tool position

12. absolute

13. MCU (Machine Control Unit)

14. A. Tool movement from one point to the next does not have to follow a specific path.
   B. Permits controlled tool travel along one axis at a time.
   C. Controls machine and tool movement as the cutter moves along the programmed path. Cutting is continuous and can be in six axes simultaneously.

15. c. Six

16. d. All of the above.

17. c. Both a and b.

18. point-to-point; the cutting tool must be fed a constantly changing series of instructions

19. d. All of the above.

20. end of block (EOB)

21. For straight line cutting, drilling, and spot welding.

22. each machining sequence and machine function into a coded block of information the MCU can understand

23. Computer-aided programming reduces and simplifies the numerical calculations that the programmer must perform when programming the machining of more complex parts.

24. languages

25. The rules used for combining the vocabulary of words, numbers, and other symbols used when writing programs.

26. geometry, machining

27. Any of the following: increased productivity, reduced machining costs, extended cutter life, reduced scrap, improved work quality, greater machine utilization.

28. Coordinates

29. Evaluate individually. Refer to Section 21.8.

30. Any two of the following: high initial cost of equipment; shortage of skilled technicians to service equipment; increased maintenance costs over traditional machine tools; machine capabilities must be fully utilized.

Direct Numerical Control (DNC)

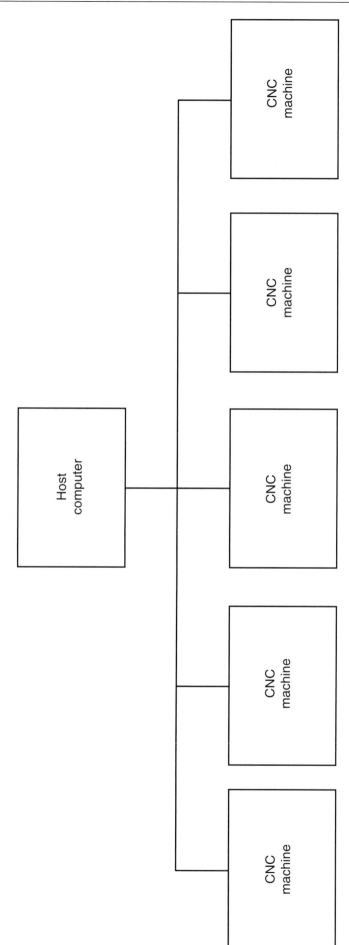

21-1

Distributed Numerical Control (DNC)

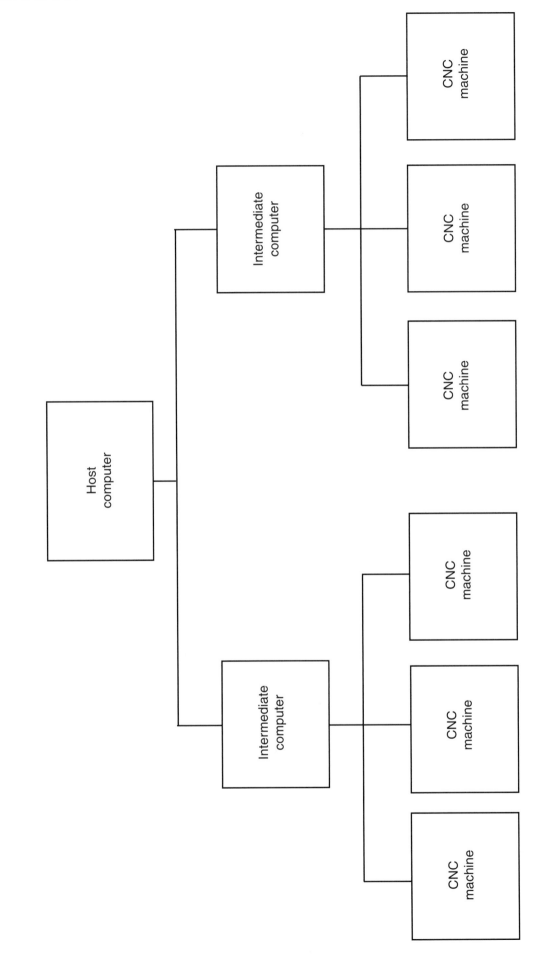

21-2

The Cartesian Coordinate System

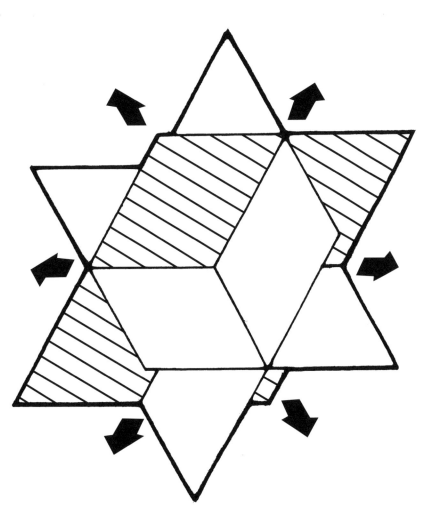

Axes of Machine Movements

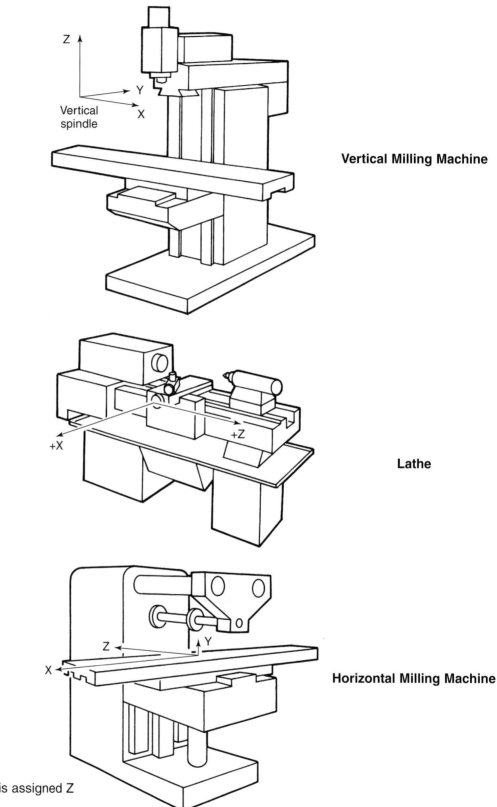

Vertical spindle

Vertical Milling Machine

Lathe

Horizontal Milling Machine

Note: Spindle motion is assigned Z axis

NC Positioning Methods

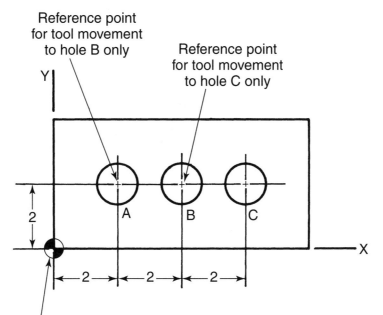

Reference point for tool movement to hole B only

Reference point for tool movement to hole C only

Zero point—tool movement only to hole A from this origin

Incremental Positioning System

Each set of coordinates has its point of origin from last point established.

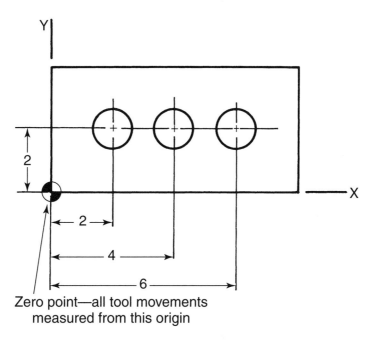

Zero point—all tool movements measured from this origin

Absolute Positioning System

In this system, all coordinates are measured from fixed point (zero point) of origin.

21-5

Contour or Continuous Machining

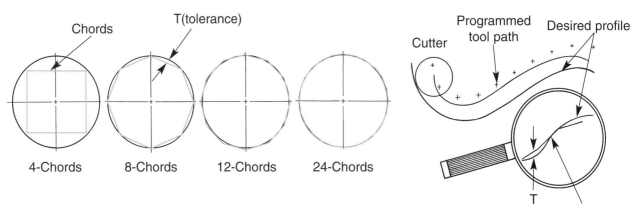

Chords   T(tolerance)

4-Chords   8-Chords   12-Chords   24-Chords

Programmed tool path segments can be as small

Cutter   Programmed tool path   Desired profile

T   Tool path

Contours obtained from contour or continuous path machining are result of a series of straight-line movements. The degree to which a contour corresponds with specified curve depends upon how many movements or chords are used. Note how, as number of chords increase, the closer the contour is to a perfect circle. The actual number of lines or points needed is determined by the tolerance allowed between design of the curved surface and one actually machined.

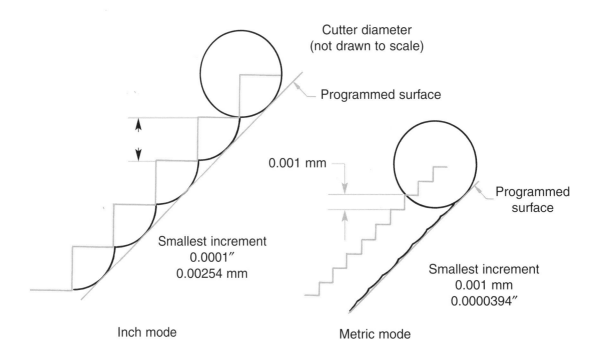

Cutter diameter
(not drawn to scale)

Programmed surface

0.001 mm

Programmed surface

Smallest increment
0.0001″
0.00254 mm

Smallest increment
0.001 mm
0.0000394″

Inch mode   Metric mode

This exaggerated illustration shows why metric machine movement increments are often preferred when contour machining. The benefit has to do with the least input increment allowed in the metric mode. In the inch mode, the least input increment is 0.0001″, which means you can input program coordinates and tool offsets down to 0.0001″. In the metric mode, the least input increment is 0.001 mm, which is less than one-half the least input increment when using the inch mode. The coordinates going into the program will then be much closer to what is desired for accurately machined parts.

21-6

Mirror Image Machining

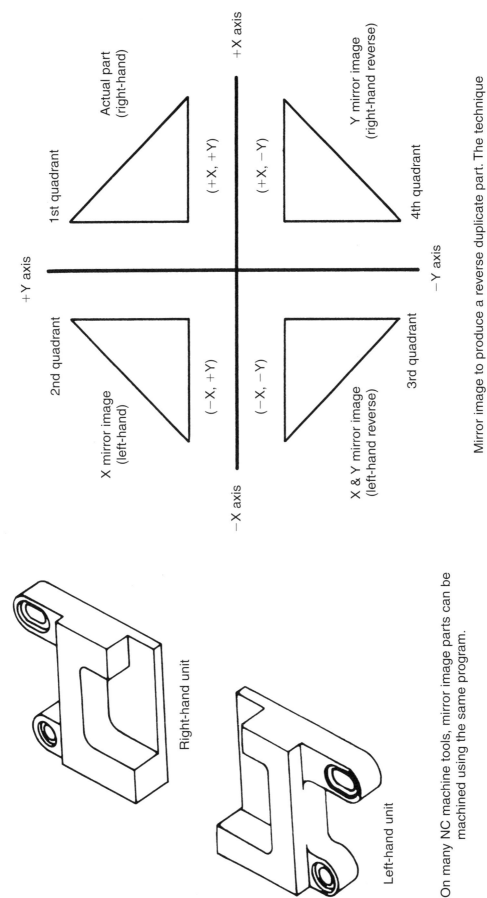

Actual part (right-hand)

1st quadrant

(+X, +Y)

Y mirror image (right-hand reverse)

(+X, −Y)

4th quadrant

+X axis

+Y axis

−Y axis

2nd quadrant

X mirror image (left-hand)

(−X, +Y)

(−X, −Y)

X & Y mirror image (left-hand reverse)

3rd quadrant

−X axis

Mirror image to produce a reverse duplicate part. The technique is known as axis inversion.

Right-hand unit

Left-hand unit

On many NC machine tools, mirror image parts can be machined using the same program.

21-7

Computer Numerical Control

Name: _____ Date: _____ Score: _____

1. Describe the differences between manual machining techniques and NC methods. _____

   _____

   _____

   _____

   _____

   _____

2. Prepare a sketch showing the Cartesian Coordinate System.

3. Prepare two similar sketches. Show incremental dimensioning on the first sketch and absolute dimensioning on the second sketch.

4. What is an NC program? _____

   _____

   _____

5. List the three basic NC systems. _____

   _____

   _____

Name _____

6. Draw sketches showing how the three NC systems differ.

7. Which of the three NC methods require the use of a computer? Why is a computer required?

   _____

   _____

8. What do the following terms mean?

   a. MCU: _____

   _____

   b. Alphanumeric data: _____

   _____

   c. Program sheet: _____

   _____

9. How is an NC program often verified? _____

   _____

   _____

10. Briefly explain adaptive control (AC). _____

   _____

   _____

   _____

Chapter 22

Automated Manufacturing

INSTRUCTIONAL MATERIALS

Text: pages 423–434
   Test Your Knowledge Questions, page 433
Workbook: pages 123–126
Instructor's Resource: pages 293–302
   Guide for Lesson Planning
   Research and Development Ideas
   Reproducible Masters:
      22-1 Flexible Machining Cell
      22-2 Work Envelope of a Robot
      22-3 Robot Configuration
      22-4 Laminated Object Manufacturing
         Process
      22-5 The Stereolithography Process
      22-6 Test Your Knowledge Questions
   Color Transparency (Binder/CD only)

GUIDE FOR LESSON PLANNING

Have the class read and study the chapter. Review the assignment using the reproducible masters as overhead transparencies and/or handouts. Discuss the following:

- The meaning of the term *automation*.
- Flexible manufacturing systems (FMS). Use Reproducible Master 22-1.
- Definition of a robot.
- How robots are used in automation.
- Design of a robot. Use Reproducible Masters 22-2 and 22-3.
- Robotic applications.
- Safety in automated manufacturing.
- Why rapid prototyping techniques have been developed.
- The Laminated Object Manufacturing (LOM) process. Use Reproducible Master 22-4.
- The stereolithography process. Use Reproducible Master 22-5.
- Other rapid modeling techniques.
- The future of automated manufacturing.

Technical Terms

Review the terms introduced in the chapter. New terms can be assigned as a quiz, homework, or extra credit. The following list is also given at the beginning of the chapter.

computer integrated manufacturing (CIM)
flexible manufacturing system (FMS)
Fused Disposition Modeling (FDM)
Just-in-Time manufacturing (JIT)
Laminated Object Manufacturing (LOM)
manipulator
robot
smart tooling
stereolithography
work envelope

Review Questions

Assign *Test Your Knowledge* questions. Copy and distribute Reproducible Master 22-6 or have students use the questions on page 433 and write their answers on a separate sheet of paper.

Workbook Assignment

Assign Chapter 22 of the *Machining Fundamentals Workbook.*

Research and Development

Discuss the following topics in class or have students complete projects on their own.

1. The advent of automation is frequently thought of as the "Second Industrial Revolution." Develop a research paper on the *first* Industrial Revolution with emphasis on working conditions, living conditions, and wages. Compare them with the working conditions, living conditions, and wages that exist today.
2. Design a bulletin board on automation.
3. Develop a program for a CNC lathe.
4. Demonstrate the operation of a CNC lathe.

TEST YOUR KNOWLEDGE ANSWERS, Page 433

1. Automation is a system for the continuous automatic production of a product.
2. Electronically, hydraulically, mechanically, pneumatically, or in combination.
3. Any order: making, inspecting, assembly, testing, packaging.
4. a. Flexible Manufacturing System
   b. Computer Integrated and Manufacturing System
   c. Flexible Machining Cell
   d. Computer Numerical Control
   e. Computer Aided Design/Computer Aided Manufacturing
   f. Laminated Object Manufacturing
5. loaders
6. a. Smart tooling involves the use of cutting tools and work-holding devices that can be readily reconfigured to produce a variety of shapes and sizes within a given part family.
   b. JIT, or Just-in-Time, is a system that eliminates the need for large inventories of materials and parts. They are scheduled for arrival at the time needed and not before.
   c. Robots are programmable, multifunctional manipulators designed to move material, parts, tools, or specialized devices through variable programmed motions for the performance of a variety of tasks.
7. The work envelope of a robot is the volume of space defined by the reach of the robot arm in three-dimensional space.
8. Student answers will vary but may include three of the following: to work in hazardous and harsh environments; perform tedious operations; for precision operations; handling heavy materials.
9. Student answers will vary but may include three of the following: Laminated Object Manufacturing (LOM); Stereolithography; Fused Disposition Modeling (FDM); Direct Shell Production Casting (DSPC).
10. laser

WORKBOOK ANSWERS, Pages 123–126

1. d. All of the above.
2. Flexible Machining Cell (FMC)
3. d. All of the above.
4. immediately
5. Parts and materials are scheduled for arrival at the time needed and not before.
6. Production can be reduced or stopped by weather delays or strikes.
7. A. Performs computations for controlling the movement of the arm and wrist to the proper location.
   B. May be hydraulic, pneumatic, or electric.
   C. The articulated arm of the robot. The end of the arm is fitted with a wrist capable of angular and/or rotational motion.
   D. Device attached to the robot wrist for specific applications, such as a gripper, welding head, or spray gun.
8. Evaluate individually. Refer to Sections 22.4.2 and 22.4.3.
9. Evaluate individually. Refer to Section 22.4.1.
10. d. All of the above.

Flexible Machining Cell

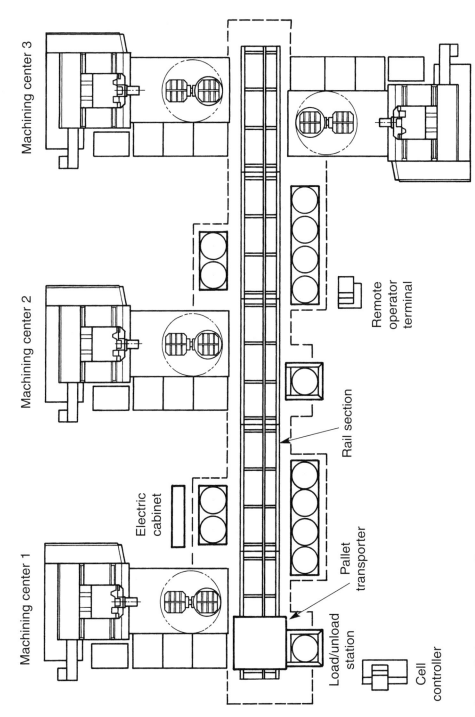

Flexible manufacturing cell that uses a pallet transporter to link the machines. A cell controller automatically queues work for immediate delivery to the next machine available.

Work Envelope of a Robot

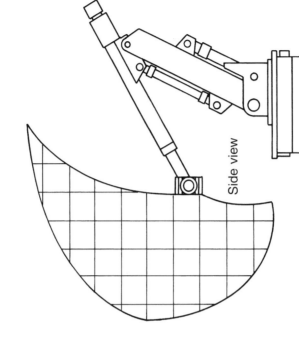

Side view

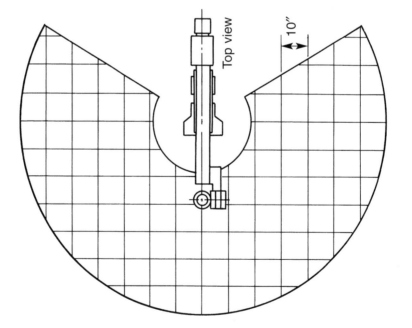

Top view

10″

The work envelope of a robot is the volume of space defined by the reach of the robot arm in three dimensions.

Robot Configuration

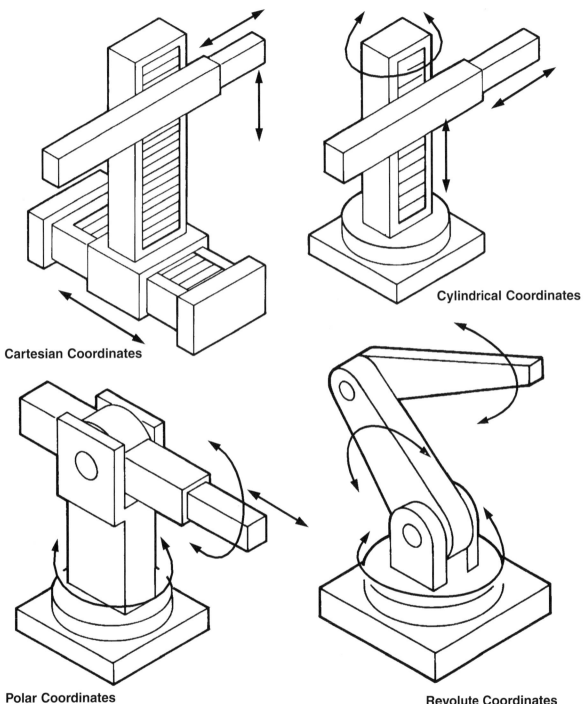

Cartesian Coordinates

Cylindrical Coordinates

Polar Coordinates

Revolute Coordinates

22-3

The Stereolithography Process

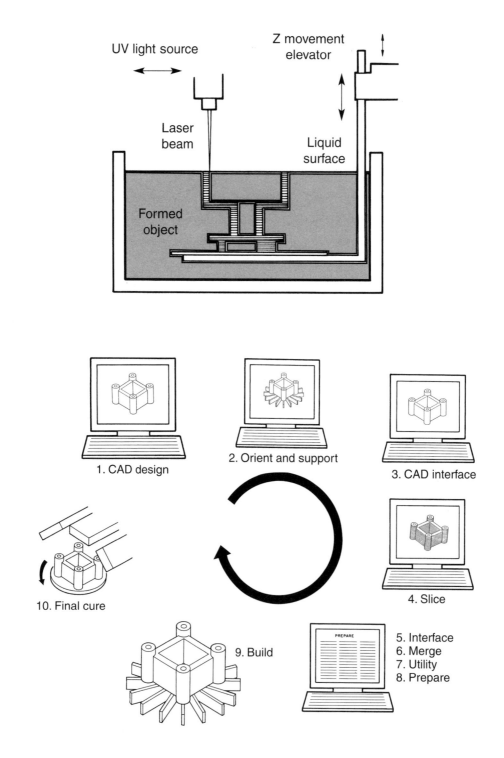

UV light source

Z movement elevator

Laser beam

Liquid surface

Formed object

1. CAD design

2. Orient and support

3. CAD interface

4. Slice

5. Interface
6. Merge
7. Utility
8. Prepare

9. Build

10. Final cure

PREPARE

22-5

Laminated Object Manufacturing Process

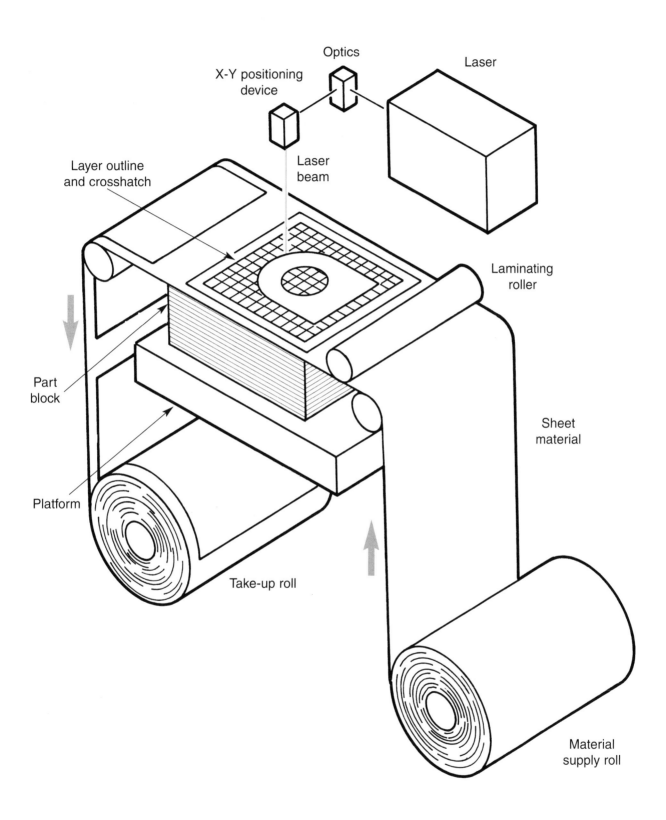

22-4

Automated Manufacturing

Name: _____ Date: _____ Score: _____

1. What is automation? _____

   _____

   _____

2. How are the machines in automation activated? _____

   _____

   _____

3. List the five basic manufacturing processes that automated machines can perform.

   _____

   _____

   _____

   _____

   _____

4. What do the following acronyms stand for in their relation to automation? (An acronym is a word formed using the initial letters of words in a phrase. For example: RPM stands for revolutions per minute.)

   a. FMS: _____

   b. CIM: _____

   c. FMC: _____

   d. CNC: _____

   e. CAD/CAM:_____

   f. LOM: _____

5. In a flexible machining cell, specially designed _____ are   5. _____
   often used to place workpieces in a machine.

6. Explain the following terms:

   a. Smart tooling: _____

   _____

   b. JIT:_____

   _____

   c. Robot: _____

   _____

   _____

Name: _____

7. In robotics, what is meant by the term "work envelope"? _____

_____

_____

8. List three current uses of the robot. _____

_____

_____

_____

9. List three techniques that are now available for the rapid prototyping of a CAD design.

_____

_____

_____

10. In several of the rapid prototyping systems, a _____ is   10. _____
    used to harden a thin layer of material as the object is
    built up.

Chapter 23

Quality Control

INSTRUCTIONAL MATERIALS

Text: pages 435–450
    Test Your Knowledge Questions, page 449
Workbook: pages 127–130
Instructor's Resource: pages 303–314
    Guide for Lesson Planning
    Research and Development Ideas
    Reproducible Masters:
        23-1 Radiographic (X-ray) Inspection
        23-2 Magnetic Particle Inspection
        23-3 Ultrasonic Inspection (*basic operation*)
        23-4 Ultrasonic Inspection (*liquid coupling/immersion type*)
        23-5 Ultrasonic Inspection (*cathode ray tube [CRT]*)
        23-6 Test Your Knowledge Questions
    Color Transparency (Binder/CD only)

GUIDE FOR LESSON PLANNING

Display a selection of quality control equipment for student/trainee inspection.

Have students/trainees read and study the chapter. Review the assignment using the reproducible masters as overhead transparencies and/or handouts. Discuss the following:

- The reason for quality control.
- The ultimate goal of quality control.
- History of quality control.
- The two basic types of quality control.
- Why destructive quality control testing may be used.
- Nondestructive quality control techniques.
- Why different types of testing are necessary.
- Measuring techniques.
- The coordinated measuring machine and its capability.
- Radiographic (X-ray) inspection and its advantages and disadvantages. Use Reproducible Master 23-1.
- Magnetic particle inspection and its advantages and disadvantages. Use Reproducible Master 23-2.
- Fluorescent penetrant inspection.
- Ultrasonic inspection. Use Reproducible Masters 23-3, 23-4, and 23-5.
- Inspection using lasers.
- Eddy-current inspection.
- Other quality control techniques.

Technical Terms

Review the terms introduced in the chapter. New terms can be assigned as a quiz, homework, or extra credit. The following list is also given at the beginning of the chapter.

coordinate measuring machine (CMM)
eddy-current test
fluorescent penetrant inspection
magnetic particle inspection
megahertz (MHz)
optical comparator
profilometer
statistical process control (SPC)
ultrasonic testing
ultraviolet light

Review Questions

Assign *Test Your Knowledge* questions. Copy and distribute Reproducible Master 23-6 or have students use the questions on page 449 and write their answers on a separate sheet of paper.

Workbook Assignment

Assign Chapter 23 of the *Machining Fundamentals Workbook.*

Research and Development

Discuss the following topics in class or have students complete projects on their own.

1. Devise a way to use an overhead projector to demonstrate the optical comparator.

2. Develop simple measuring fixtures to check a simple project against its plans.

3. Penetrants described in the text are easy to use. The cost is within reach of most budgets. Carefully analyze the needs of your shop. If needed, present your analysis to request that penetrant materials be purchased.

4. Select a machine part and examine it carefully. What points on the piece come under the quality control program in the plant where it was manufactured? What points must be checked against specifications if the part is to be interchangeable with other components of the assembly?

5. Devise a quality control program for your training area.

6. Arrange a field trip to a manufacturing plant. Request a demonstration of the Magnaflux technique.

7. Check the micrometers and Vernier measuring tools in your instructional area. Make any needed repairs and adjustments.

8. Show a film or video on quality control. Preview it and prepare an outline and quiz on the film's more significant points of interest.

TEST YOUR KNOWLEDGE ANSWERS, Page 449

1. d. All of the above.
2. Destructive testing, the part is destroyed during test.
   Nondestructive testing, the part can be used after test.
3. precision tool calibration
4. X-rays
5. Any order: inspection sensitivity is high; image produced is geometrically accurate; a permanent record is produced; image interpretation is highly accurate.
6. magnafluxing
7. Evaluate individually. Refer to Section 23.3.4.
8. b. It guarantees that the parts being produced meet standards and specifications.
9. optical comparator
10. A magnetic field is set up electrically within the part. Fine iron particles are blown or flowed in liquid suspension on the part. They will outline the flaw.
11. ferrous
12. b. A high-frequency sound beam.
13. Evaluate individually. Refer to Figure 23-29 in the text.
14. Pulse echo. Uses sound waves generated by a transducer that travels through the work. The reflected sound waves (echoes) locate the flaws. One crystal is used to both transmit the sound and receive the echoes.
    Through inspection. Has one crystal that transmits the waves through the piece, and another crystal picks up the signal on the opposite end of the piece.
15. Evaluate individually. Refer to Section 23.3.6.

WORKBOOK ANSWERS, Pages 127–130

1. prevent
2. d. All of the above.

3. Part is destroyed during testing. Specimen is selected at random and gives no assurance that defective parts could slip by.

4. Evaluate individually.

5. Usefulness of part is not impaired. Each piece is tested individually and as a part of a complete assembly.

6. It guarantees measuring tool accuracy by checking them against known standards.

7. electronically

8. An enlarged image of the part being inspected is projected onto a screen where it is superimposed upon an accurate drawing overlay of the part. Very small size variations can be noted by skilled operator.

9. automates

10. Statistical process control involves measuring a mathematically selected number of parts in a production run.

11. Passing gamma rays through the part and onto light sensitive film to detect flaws. The developed film has an image of the internal structure of the part.

12. d. All of the above.

13. To detect flaws on or near the surface of ferromagnetic (iron-based) metals.

14. Cannot find flaws in nonferrous materials and it only shows serious defects.

15. A penetrant solution is applied to the surface of the part. Capillary action pulls the penetrant into the flaw. The surface is rinsed clean and a developer applied. When inspected under ultraviolet light, flaws will glow with fluorescent brilliance.

16. By coating the part with a red liquid dye which soaks into any flaw. After washing off the dye, a developer is dusted or sprayed on the part. Flaws show up as red against the white of the developer.

17. sound waves

18. d. All of the above.

19. It is based on the fact that flaws in a metal product will cause impedance changes in a coil brought near it. Different eddy-currents will result in test coils placed next to metal parts with and without flaws. This difference determines which parts pass or fail inspection.

20. c. variations in dimension of a metal product

Radiographic (X-ray) Inspection

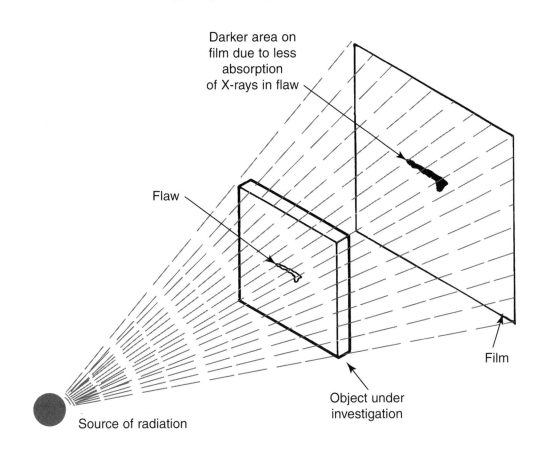

Darker area on
film due to less
absorption
of X-rays in flaw

Flaw

Film

Object under
investigation

Source of radiation

How Radiographic Inspection Works

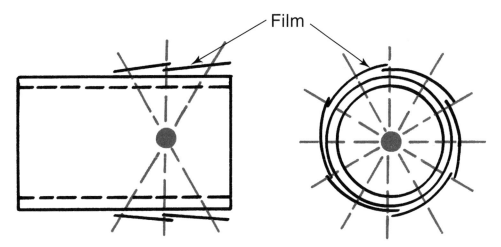

Film

Inspecting Cylindrical Objects

A flaw causes more exposure of the film, so an image of the flaw is shown on the
film when it is developed.

23-1

Magnetic Particle Inspection

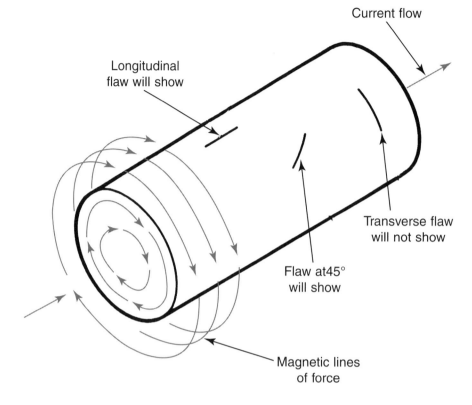

Theory, scope, and limitations of magnetic particle inspection technique.

Iron particles

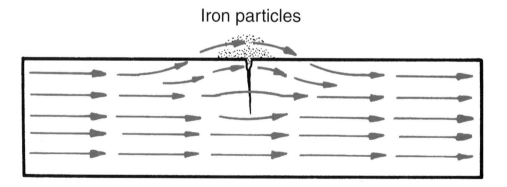

Crack in steel bar generates a magnetic field outside the part to hold on iron particles. Buildup of iron particles makes even tiny flaws visible.

Ultrasonic Inspection

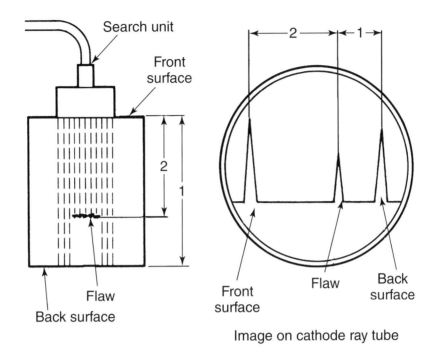

How ultrasonic sound waves are used to detect and locate a flaw in a test piece.

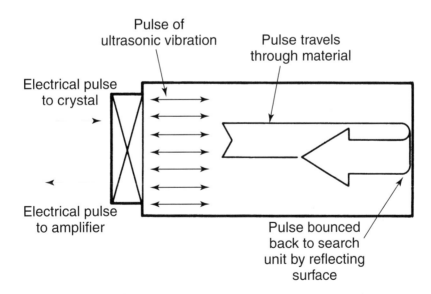

How sound waves travel through a part and bounce back to locate flaws in the material.

23-3

Ultrasonic Inspection

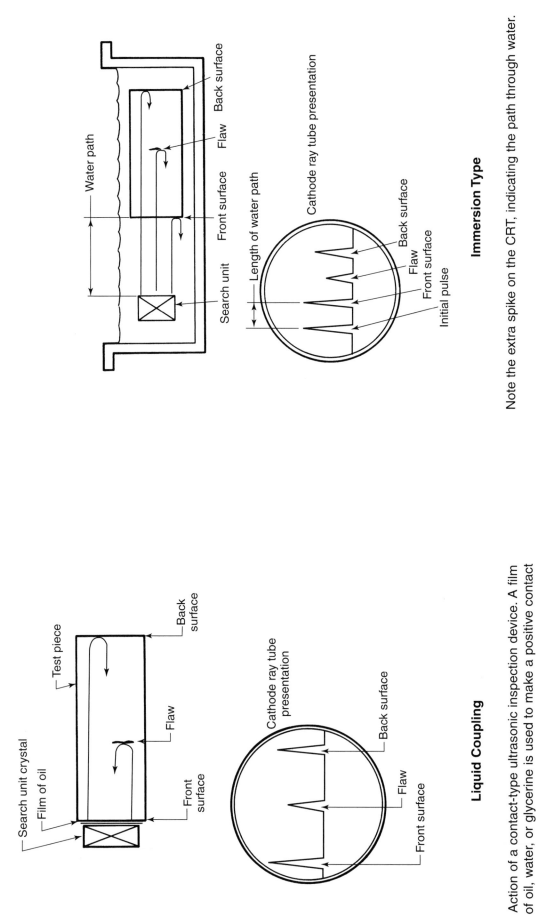

23-4

Immersion Type

Note the extra spike on the CRT, indicating the path through water.

Liquid Coupling

Action of a contact-type ultrasonic inspection device. A film of oil, water, or glycerine is used to make a positive contact between the transducer and the test piece.

Ultrasonic Inspection

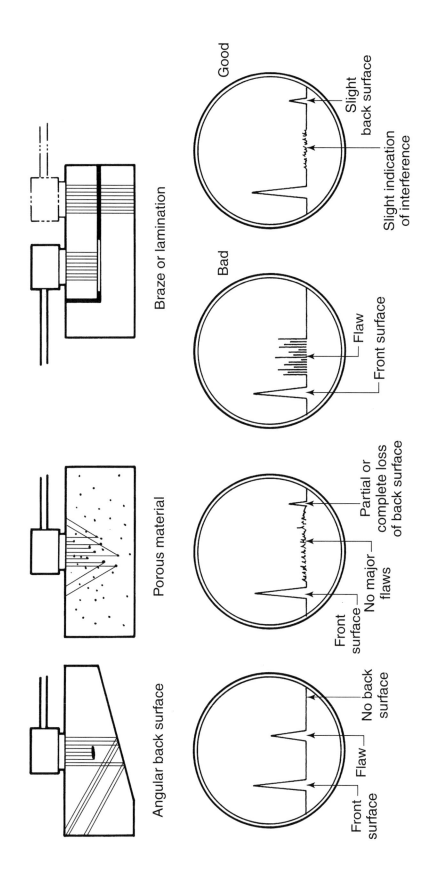

Braze or lamination

Porous material

Angular back surface

Good

Slight back surface

Slight indication of interference

Bad

Flaw

Front surface

Partial or complete loss of back surface

Front surface

No major flaws

No back surface

Flaw

Front surface

Quality Control

Name: _____Date: _____ Score: _____

1. Quality control is an important segment of industry.   1. _____
   Its purpose is to:
   a. Improve product quality.
   b. Maintain quality.
   c. Help to reduce costs.
   d. All of the above.
   e. None of the above.

2. Quality control falls into two basic classifications. Name and explain each. _____

   _____

   _____

   _____

   _____

3. Precision measuring tools, such as micrometers, Vernier   3. _____
   tools, or dial indicators, are inspected and calibrated in a
   _____ laboratory.

4. Inspection by radiography involves the use of _____ and   4. _____
   gamma radiation.

5. List four advantages of the radiographic inspection process.

   _____

   _____

   _____

   _____

6. Magnetic particle inspection is commonly known as   6. _____
   _____.

7. Describe the fluorescent penetrant inspection process. _____

   _____

   _____

   _____

8. Quality control is an important industrial tool because: 8. _____
   a. It can be done easily.
   b. It guarantees that the parts being produced meet standards and specifications.
   c. It can be done by unskilled labor.
   d. All of the above.
   e. None of the above.

23-6

(continued)

Name: _____

9. The _____ is an optical gaging instrument designed for   9. _____
the inspection of small parts and sections of larger parts.

10. Explain how the magnetic particle inspection technique operates. _____

_____

_____

_____

_____

11. Only _____ metals can be inspected by the magnetic   11. _____
particle technique.

12. Ultrasonic inspection makes use of: 12. _____
    a. Accurately made measuring fixtures.
    b. A high-frequency sound beam.
    c. X-rays.
    d. All of the above.
    e. None of the above.

13. *In the space below or on a separate sheet of paper,* make a sketch showing the two methods of liquid coupling used for ultrasonic testing.

23-6

(continued)

Name: _____

14. List the two basic categories of ultrasonic testing. Briefly describe each._____

    _____

    _____

    _____

    _____

15. *In the space below or on a separate sheet of paper,* make a sketch showing how ultrasonic testing is done.

Chapter 24

Metal Characteristics

LEARNING OBJECTIVES

After studying this chapter, students will be able to:
○ Explain how metals are classified.
○ Describe the characteristics of metals.
○ Recognize the hazards that are posed when certain metals are machined.
○ Explain the characteristics of some reinforced composite materials.

INSTRUCTIONAL MATERIALS

Text: pages 451–466
    Test Your Knowledge Questions, page 465
Workbook: pages 131–134
Instructor's Resource: pages 315–322
    Guide for Lesson Planning
    Research and Development Ideas
    Reproducible Masters:
        24-1 Tungsten Carbide Tool Coating
        24-2 Test Your Knowledge Questions
    Color Transparency (Binder/CD only)

GUIDE FOR LESSON PLANNING

Before studying the chapter, have the class define the term *metal* in their own words. Is there a definition that will describe all metals? Have samples of as many different metals as possible. They should be identical in size. Permit the class to handle them, feel their weights (the weights of two aluminum alloys will differ slightly), and note how the metals differ in color.

Have the class read and study all or part of the chapter. Review the assignment and discuss the following:

- The classification of metals.
- Determining whether a metal is a ferrous or a nonferrous metal.
- Identifying an alloy from a base metal.
- Types of ferrous metals.
- Alloying elements added to ferrous metals to improve their properties.
- Why coatings are put on tungsten carbide cutting tools. Use Reproducible Master 24-1.
- Stainless steels.
- The AISA/SAE codes for identifying the physical characteristics of steel.
- Color coding of steel.
- The spark test for determining the grade of steel.
- Nonferrous metals.
- The metal known as aluminum.
- Magnesium and the precautions that must be taken when machining it.
- Titanium and how it is machined.
- Copper-based alloys.
- Precautions that must be taken when machining beryllium copper.

- High temperature metals.
- Rare metals.
- Honeycomb and composites.
- How to handle metal safely in the shop/lab.

Technical Terms

Review the terms introduced in the chapter. New terms can be assigned as a quiz, homework, or extra credit. The following list is also given at the beginning of the chapter.

alloy
Aluminum Association Designation System
base metal
carbon content
ductility
ferrous
honeycomb
nonferrous
red hardness
tungsten carbide

Review Questions

Assign *Test Your Knowledge* questions. Copy and distribute Reproducible Master 24-2 or have students use the questions on page 465 and write their answers on a separate sheet of paper.

Workbook Assignment

Assign Chapter 24 of the *Machining Fundamentals Workbook.*

Research and Development

Discuss the following topics in class or have students complete projects on their own.

1. Prepare a display showing specimens of ferrous, nonferrous, high temperature, and rare metals. Label them according to type, classification, and use.
2. Many terms are used to describe the various properties of metals. Research the meanings of the following terms as they pertain to metals.
    Machinability Ductility
    Malleability Elasticity
    Brittleness Hardness
    Toughness Tensile Strength
    Yield Point Elongation
    Stress Plasticity
3. Secure samples of honeycomb. Prepare a report explaining why it is difficult to

machine by conventional methods. Include some of the techniques employed to machine the material.
4. Secure samples of composites. Prepare a report on how they are made and shaped.
5. Contact firms manufacturing high-temperature metals and request printed material as well as photographs for your shop technical library.
6. Show a film or video on how aluminum is made. Discuss the details of the film with the class.
7. Show a film or video on steel making. Prepare a quiz to be given after the presentation and discussion.
8. Secure copies of trade magazines dealing with metals and other materials.

TEST YOUR KNOWLEDGE ANSWERS, Page 465

1. Any order: ferrous metals, nonferrous metals, high temperature metals, rare metals.
2. ferrous
3. b. Cast iron has a hard surface scale.
4. iron, carbon
5. By the amount of carbon they contain measured in percentage or in points.
6. black oxide scale
7. sulphur, lead
8. harder, stronger, tougher
9. alloy
10. corrosion
11. Any order: austenitic, martensitic, ferritic.
12. nonferrous
13. lightest
14. strong, half, heavy
15. copper
16. zinc
17. tin
18. The fine dust generated by machining and filing can cause severe respiratory damage.
19. high-temperature
20. Honeycomb structures give existing metals greater strength and rigidity while reducing weight. Sections of thin material (aluminum, stainless steel, titanium, and nonmetals like Nomex® fabric) are bonded together to form a structure that is similar in appearance to the wax comb that bees create to store honey.

21. Composites use fibers of both conventional and uncommon materials, in both pure and alloy forms. Fibers such as pure iron, graphite, boron, and fiberglass are bonded together in a special plastic matrix under heat and pressure. These materials are generally lighter, stronger, and more rigid than many conventional metals.

WORKBOOK ANSWERS,
Pages 131–134

1. ferrous
2. A metal that contains no alloying metals.
3. A mixture of two or more metals.
4. b. 0.30, 30
5. d. All of the above.
6. b. smooth bright finish
7. c. 0.60 to 1.50, 60 to 150
8. d. All of the above.
9. They are more costly to produce because of the increased number of special operations that must be performed in their manufacture.
10. c. Chromium
11. b. Manganese
12. a. Nickel
13. Steels found in devices that are used to cut, shear, or form materials. It may be either carbon or alloy steel. Steels in the lower carbon range are used for tools subject to shock. Steels in the higher carbon range are used when tools with keen cutting edges are required.
14. It is the hardest human-made metal.
15. Magnesium
16. Evaluate individually. Refer to Section 24.2.7.
17. e. All of the above.
18. nonferrous
19. You must know what alloy is used because each alloy requires a different procedure.
20. It indicates the degree of hardness of an alloy.
21. They are extremely strong and corrosion-resistant under most conditions; the alloys are lighter than most commercially available metals; they can be shaped and formed easily; they are readily available in a multitude of sizes, shapes, alloys, and tempers.
22. Magnesium chips are highly flammable and burn at very high temperatures. It is a low-level radioactive material.
23. Water-base cutting fluids.
24. A respirator-type face mask must be worn; special procedures must be followed when cleaning machines used to machine the alloy.
25. Student answers will vary but may include two of the following: strong as steel, weighs only half as much as steel, extremely resistant to corrosion, most titanium alloys are capable of continuous use at temperatures up to about 800°F.
26. Metal with high strength for extended periods of time at elevated temperatures.
27. c. Tungsten
28. It has a very high strength-to-weight ratio and rigidity-to-weight ratio.
29. They are generally lighter, stronger, and more rigid than many conventional metals.
30. Different types of fibers are bonded together in a plastic matrix under heat and pressure.

Tungsten Carbide Tool Coating

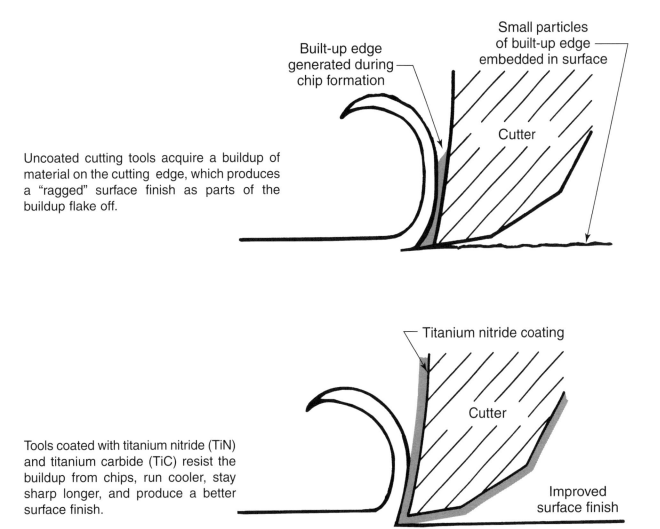

Built-up edge generated during chip formation

Small particles of built-up edge embedded in surface

Cutter

Uncoated cutting tools acquire a buildup of material on the cutting edge, which produces a "ragged" surface finish as parts of the buildup flake off.

Titanium nitride coating

Cutter

Tools coated with titanium nitride (TiN) and titanium carbide (TiC) resist the buildup from chips, run cooler, stay sharp longer, and produce a better surface finish.

Improved surface finish

Metal Characteristics

Name: _____ Date: _____ Score: _____

1. What are the four main categories of metals? _____

   _____

   _____

2. Iron and steel are classified as _____ metals.   2. _____

3. Carbide cutting tools are recommended for machining   3. _____
   cast iron because:
   a. Cast iron is hard and brittle.
   b. Cast iron has a hard surface scale.
   c. Cast iron is difficult to machine.
   d. All of the above.
   e. None of the above.

4. Carbon steel is an alloy of _____ and _____.   4. _____

   _____

5. How are carbon steels classified? _____

   _____

6. Hot rolled steel is characterized by the _____ on its surface.   6. _____

7. The machinability of carbon steel is improved if _____ or   7. _____
   _____ is added as an alloying element.
   _____

8. Nickel, chromium, molybdenum, vanadium, and tung-   8. _____
   sten are used to make steel _____, _____, and _____.
   _____

   _____

9. Drills, reamers, some milling cutters, and similar tools   9. _____
   are usually made from _____ steel.

10. The chief characteristic of stainless steel is its resistance   10. _____
    to _____.

11. List the three basic groups of stainless steel. _____

    _____

12. Aluminum, magnesium, and titanium are _____ metals.   12. _____

13. Magnesium is the _____ of the structural metals.   13. _____

14. Titanium is a metal that is as _____ as steel but only   14. _____
    _____ as _____.
    _____

    _____

Name: _____

15. Brass and bronze are _____-based alloys.   15. _____

16. Brass is an alloy of copper and _____.   16. _____

17. Bronze is an alloy of copper and _____.   17. _____

18. Why must a machinist take special precautions when working beryllium copper?_____

_____

_____

_____

_____

_____

19. Nickel-based alloys, molybdenum, tantalum, and tungsten   19. _____
    are classified as _____ metals.

20. What is the structural material known as honeycomb?_____

_____

_____

_____

_____

_____

21. What are composites?

_____

_____

_____

_____

_____

Chapter 25

Heat Treatment of Metals

LEARNING OBJECTIVES

After studying this chapter, students will be able to:
- ○ Explain why some metals are heat-treated.
- ○ List some of the metals that can be heat-treated.
- ○ Describe some types of heat-treating techniques and how they are performed.
- ○ Case harden low-carbon steel.
- ○ Harden and temper some carbon steels.
- ○ Compare hardness testing techniques.
- ○ Point out the safety precautions that must be observed when heat-treating metals.

INSTRUCTIONAL MATERIALS

Text: pages 467–488
   Test Your Knowledge Questions, page 487
Workbook: pages 135–138
Instructor's Resource: pages 323–330
   Guide for Lesson Planning
   Research and Development Ideas
   Reproducible Masters:
      25-1 Critical Range Diagram for Plain
         Carbon Steel
      25-2 Test Your Knowledge Questions
   Color Transparencies (Binder/CD only)

GUIDE FOR LESSON PLANNING

Prepare a number of center punches in various stages of heat treatment. The punch in the annealed stage should show a blunted point because it is not hard enough. The fractured pieces of a glass hard punch will illustrate what happens when steel is hardened but not tempered. A properly heat-treated punch will show no signs of wear.

Have the class read and study the chapter, paying particular attention to the illustrations. Review the assignment and discuss the following:
- Everyday items they believe must be heat treated to fulfill their function.
- What heat treatment involves.
- Metals that can be heat treated.
- Types of heat treatment.
- Purpose of stress-relieving.
- The annealing process.
- The process of normalizing steel.
- The hardening process and why additional heat treatment is usually required.
- Surface hardening and the various techniques used.
- Why case hardening is used and how it is done.
- Why tempering or drawing is done.
- Heat treatment of metals other than steel.

- Heat-treatment equipment.
- Quenching media and its uses.
- How carbon steel is hardened and tempered.
- How low-carbon steel is case hardened.
- Hardness testing and why it is performed.
- The equipment used for hardness testing and how each type is used.
- Heat-treating safety.

Emphasize the safety precautions that must be observed when performing heat-treating operations.

Demonstrate the following:
- Annealing.
- Hardening and tempering.
- Case hardening techniques.
- Furnace operation.
- The use of the various testing devices.

When demonstrating heat-treating techniques, be sure there is adequate ventilation and all safety precautions are observed by the instructor and students/trainees.

Plan work allowing the class to develop basic heat-treating skills.

Review the demonstrations and provide students with the opportunity to ask questions.

Technical Terms

Review the terms introduced in the chapter. New terms can be assigned as a quiz, homework, or extra credit. The following list is also given at the beginning of the chapter.

> *annealing*
> *Brinell hardness tester*
> *case hardening*
> *hardness number*
> *normalizing*
> *Rockwell hardness tester*
> *scleroscope*
> *stress-relieving*
> *tempering*
> *Webster hardness tester*

Review Questions

Assign *Test Your Knowledge* questions. Copy and distribute Reproducible Master 25-2 or have students use the questions on page 487 and write their answers on a separate sheet of paper.

Workbook Assignment

Assign Chapter 25 of the *Machining Fundamentals Workbook*.

Research and Development

Discuss the following topics in class or have students complete projects on their own.

1. Prepare a glossary of heat treating terms. Duplicate and distribute copies to the class.
2. When heat-treating aluminum alloys, the terms *solution heat treatment* and *precipitation hardening* are used. What do they mean?
3. The Metcalf Experiment is one method used to show the grain structure of heat-treated steel and the effects caused by overheating. How is it performed? Perform the experiment and mount the pieces that show the results on a panel for observation.
4. Demonstrate the proper way to harden and temper a piece of tool steel.
5. Demonstrate the proper way to case harden low-carbon steel by carburizing. Use Kasenit as the carbon source.
6. Secure samples of work that have been heat-treated by various techniques.
7. The Moh Scale was the first hardness testing technique. Research the Moh Scale and explain how it was used.
8. Secure handbooks from the various steel manufacturers and/or distributors for inclusion in the shop's technical library.
9. Arrange a field trip to a local industry that has a heat-treating area. Ask to have the various hardness testers demonstrated.
10. Secure literature on various hardness testers for inclusion in shop's technical library.

TEST YOUR KNOWLEDGE ANSWERS, Page 487

1. d. All of the above.
2. It is the controlled heating and cooling of a metal to obtain certain desirable changes in its physical properties.
3. Students may give an example such as: 60 point carbon steel would contain 60/100 (0.60) of 1% carbon. Evaluate individually.
4. Student answers will vary but may include four of the following: magnesium, many aluminum alloys, copper, titanium, beryllium copper.
5. oil, brine, air, nitrogen
6. c. Done to reduce stress that has developed in parts that have been welded, machined, or cold worked during processing.

7. a. Involves heating metal to slightly above its upper critical temperature and then permitting it to cool slowly in insulating material. Hardness of the metal is reduced.

8. b. Used to refine grain structure of steel and to improve its machinability.

9. e. Only the outer surface of low-carbon steel is hardened while the inner portion remains relatively soft and tough.

10. d. Used when only a medium-hard surface is required on high-carbon or alloy steels.

11. f. Accomplished by heating metal to its critical range and cooling rapidly.

12. c. Tough.

13. It is quieter, does not have to be ventilated, temperature can be more accurately controlled, and it is safer to operate.

14. pyrometer

15. Test used to accurately measure the degree of the hardness/softness of a metal compared to known standards.

16. The Brinell Hardness System, Rockwell Hardness System, and the Shore Scleroscope Hardness Tester.

17. Evaluate individually. Refer to Section 25.5.

18. Evaluate individually. Refer to Section 25.9.

WORKBOOK ANSWERS, PAGES 135–138

1. Improved resistance to shock, toughness development, and increased wear resistance and hardness.

2. 50

3. Removes internal stresses that have developed in parts that have been cold worked, machined, or welded. Also known as stress-relieving.

4. The part is placed in a metal box and the entire unit is heated, then allowed to cool slowly in the sealed furnace. It is used to prevent the work from scaling or decarbonizing.

5. Annealing reduces the hardness of a metal, making it easier to machine or work.

6. Normalizing is a process employed to refine the grain structure of some steels thereby improving machinability.

7. Surface hardening is used when only a medium hard surface is required on high-carbon or alloy steels.

8. Hardening is accomplished by heating metal to its critical range and cooling rapidly. It is normally employed to obtain optimum physical qualities in steel.

9. The temperature at which steel will harden. It ranges from 1400°F–2400°F (760°C–1316°C), depending on the alloy and carbon content.

10. b. Flame

11. Carburizing, liquid salt or cyanide, nitriding or gas method.

12. Case hardening puts a hard shell on the surface of low-carbon steel while the inner portion of the metal remains soft.

13. carburizing

14. Tempering or drawing is used to reduce a metal's brittleness or hardness. It involves heating the steel to below the metal's critical range. It is held at the temperature until penetration is complete and then it is quenched.

15. d. All of the above.

16. titanium

17. d. All of the above.

18. sealed and a vacuum drawn to remove atmospheric gases that might contaminate the metal being heat-treated

19. With the use of a color chart, the temperature can be judged by the color of the metal as it heats.

20. Evaluate individually. Students may give one of two methods listed in Section 25.7.

21. Hardness testing

22. They use the technique that measures the distance a steel ball or special-shaped diamond penetrates into the metal under a specific load.

23. The scleroscope uses a technique that drops a small diamond hammer onto the test piece and the height of the rebound of the hammer is used to determine the hardness of the metal.

24. a. Webster hardness tester

25. Evaluate individually. Refer to Section 25.9.

Critical Range Diagram for Plain Carbon Steel

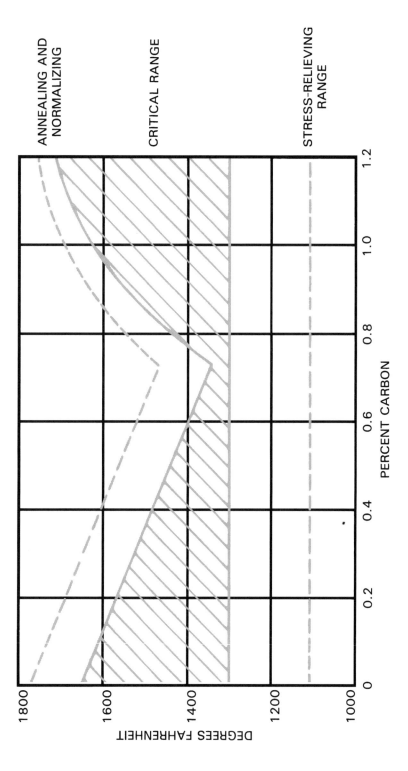

ANNEALING AND NORMALIZING

CRITICAL RANGE

STRESS-RELIEVING RANGE

PERCENT CARBON

DEGREES FAHRENHEIT

Heat Treatment of Metals

Name: _____ Date: _____ Score: _____

1. Heat-treating is done to:
   a. Obtain certain desirable changes in the metal's physical characteristics.
   b. Increase the hardness of the metal.
   c. Soften (anneal) the metal.
   d. All of the above.
   e. None of the above.

   1. _____

2. What does heat-treating involve? _____

   _____

   _____

   _____

   _____

   _____

   _____

3. Carbon steels are classified by the percentage of carbon they contain, expressed in "points" or hundredths of 1%. If this statement is true, briefly explain what it means. _____

   _____

   _____

   _____

   _____

   _____

4. Other than steel, list four other metals that are capable of being heat-treated. _____

   _____

   _____

   _____

5. In addition to water, _____, _____, blasts of cold _____, or liquid _____ may be used as a quenching medium.

   5. _____

   _____

   _____

   _____

Name: _____

- Each word in the left column matches one of the sentences. Write the letter next to the appropriate number to match the words and statements.

_____ 6. Stress-relieving.

_____ 7. Annealing.

_____ 8. Normalizing.

_____ 9. Case hardening.

_____ 10. Surface hardening.

_____ 11. Hardening.

a. Involves heating metal to slightly above its upper critical temperature and then permitting it to cool slowly in insulating material. Hardness of the metal is reduced.

b. Used to refine grain structure of steel and to improve its machinability.

c. Done to reduce stress that has developed in parts that have been welded, machined, or cold worked during processing.

d. Used when only a medium-hard surface is required on high-carbon or alloy steels.

e. Only the outer surface of low-carbon steel is hardened while the inner portion remains relatively soft and tough.

f. Accomplished by heating metal to its critical range and cooling rapidly.

12. Tempering a piece of hardened steel makes it:
    a. Brittle.
    b. Soft.
    c. Tough.
    d. All of the above.
    e. None of the above.

12. _____

13. What advantages does an electric heat-treating furnace have over a gas fired heat-treating furnace?

_____

_____

_____

14. The _____ is used to measure and monitor the high temperatures needed in heat-treating.

14. _____

15. What is hardness testing? _____

_____

_____

_____

16. List three types of commonly used hardness testers._____

_____

_____

_____

Name: _____

17. What safety precautions must be observed when lighting a gas-fired heat-treating furnace?

_____

_____

_____

_____

_____

18. List five safety precautions that must be observed when heat-treating metal.

_____

_____

_____

_____

_____

Chapter **26**

Metal Finishing

LEARNING OBJECTIVES

After studying this chapter, students will be able to:
○ Describe how the quality of a machined surface is determined.
○ Explain why the quality of a machined surface has a direct bearing on production costs.
○ Describe some metal finishing techniques.

INSTRUCTIONAL MATERIALS

Text: pages 489–502
 Test Your Knowledge Questions, page 502
Workbook: pages 139–142
Instructor's Resource: pages 331–340
 Guide for Lesson Planning
 Research and Development Ideas
 Reproducible Masters:
  26-1 Surface Condition and Values
  26-2 Lay Symbols
  26-3 Typical Surface Finishes
  26-4 Test Your Knowledge Questions
 Color Transparency (Binder/CD only)

GUIDE FOR LESSON PLANNING

A selection of products illustrating the various finishes described in this chapter should be available for class examination.

Have the class read and study the chapter. Review the assignment using the reproducible masters as overhead transparencies and/or handouts. Discuss the following:
- The definition of *surface finish.*
- Why surface roughness standards were devised.
- How to understand the symbols used to specify the finish of a machined surface.

- Degrees of surface roughness.
- How surface finish affects the economics of machined surfaces.
- Other metal finishing techniques and reasons for using them.
- Organic coatings and their application.
- Inorganic coatings and the application processes and materials involved.
- Why the chemical blackening process is used.
- Metal coatings and their application.
- Mechanical finishes and their processes of application.

Technical Terms

Review the terms introduced in the chapter. New terms can be assigned as a quiz, homework, or extra credit. The following list is also given at the beginning of the chapter.

 anodizing
 electroplating
 lay
 metal spraying
 microinches
 micrometers
 roller burnishing

surface roughness standards
vitreous enamel
waviness

Review Questions

Assign *Test Your Knowledge* questions. Copy and distribute Reproducible Master 26-4 or have students use the questions on page 502 and write their answers on a separate sheet of paper.

Workbook Assignment

Assign Chapter 26 of the *Machining Fundamentals Workbook.*

Research and Development

Discuss the following topics in class or have students complete projects on their own.

1. Secure a copy of the publication *Surface Texture*–ANSI/ASME B46.1 for the shop technical library. This ANSI publication is on the measurement of surface roughness.

2. Make a collection of brochures advertising the various types of surface roughness/texture checking equipment. Develop a bulletin board display around them.

3. Secure or make samples of machined surfaces that match the various degrees of surface roughness. Mount them on a display panel. Identify each sample with the method employed to machine it and the correct symbol of roughness.

4. Prepare a paper that will explain the techniques used to develop average roughness values. They are explained in many drafting books, machinists' handbooks, and the ANSI/ASME publication. The term RMS is often used in the formulas. What does it mean?

5. Demonstrate electroplating. Secure the necessary equipment from the science department.

6. Prepare a demonstration of the anodizing process.

7. Devise and construct a safe method to clean work made in the shop.

8. Secure samples of work that have been electroplated and anodized.

9. Contact a local machine shop and find out what equipment they use to check for surface quality.

10. Prepare a term paper on flame spraying.

TEST YOUR KNOWLEDGE ANSWERS, Page 502

1. *f*
2. Each machinist interpreted it differently because it was not based on a specific set of standards.
3. microinches, micrometers; Microinches equal millionths of an inch. Micrometers equal millionths of a meter.
4. b. Smoothly rounded undulations caused by tool and machine conditions.
5. Term used to describe the direction of the predominate tool marks.
6. rougher
7. Evaluate individually. Refer to Section 26.2.
8. cleaned
9. organic
10. Brushing, spraying, roller coating, dipping, and flow coating.
11. Ordinary anodizing, hardcoat anodizing, and electrobrightening.
12. Metal is deposited electrically on the desired surface.

WORKBOOK ANSWERS, Pages 139–142

1. Each machinist interpreted specifications differently and pieces were often better finished than necessary, raising costs.
2. d. All of the above.
3. d. All of the above.
4. irregularities
5. roughness gage, profilometer
6. The smoothly rounded peaks and valleys caused by tool vibration and chatter.
7. b. direction of predominate tool marks, grain, or pattern of surface roughness
8. higher
9. Evaluate individually. Refer to Section 26.2.
10. Lapping
11. Paints, varnishes, lacquers, enamels, various plastic-base materials, epoxies.
12. Anodizing forms a protective layer of aluminum oxide on aluminum parts. The three anodizing classes are ordinary anodizing, hard coat anodizing, and electrobrightening.
13. Electrobrightening
14. F. Buffing

15. G. Power brushing
16. A. Anodizing
17. B. Vitreous enamel
18. C. Chemical blackening
19. D. Electroplating
20. E. Metal spraying
21. It is applied as a powder (frit), or as a thin slurry known as slip. After the finish dries, the material is fired at about 1500°F (815°C) until it fuses to the metal surface.
22. Any of the following: enhances appearance of part; protects machined surface against humidity and corrosion; reduces glare; abrasive resistance is improved; adhesion qualities are improved.
23. Builds up worn or scored surfaces so they can be remachined to required size, and superhard coatings can be applied when abrasion-resistant surfaces are needed.
24. A process for depositing a metallic coating on a workpiece. It uses a water-cooled barrel several feet long and about one inch in diameter that is fitted with valving for introducing gases and material to be sprayed.
25. Student answers will vary but may include the following: it can be fully automated; it can be used to apply coatings with high melting points to fully heat-treated parts without danger of changing the metallurgical properties or strength of the part and without danger of thermal distortion; and almost any material that can be melted without decomposing can be sprayed.

Surface Condition and Values

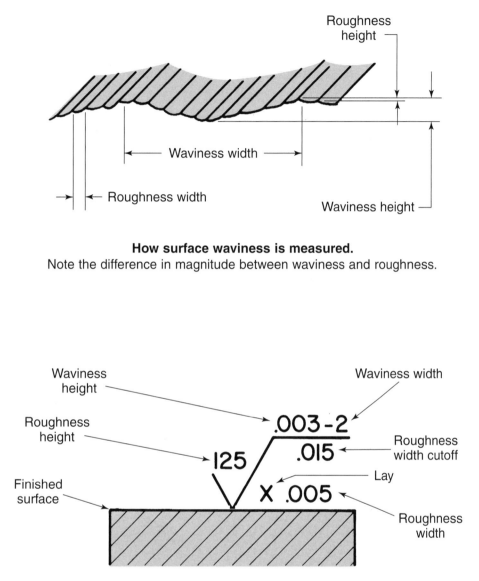

How surface waviness is measured.
Note the difference in magnitude between waviness and roughness.

On drawings, symbols and numbers show roughness, waviness, and lay.
They specify finishes required on a surface.

26-1

Lay Symbols

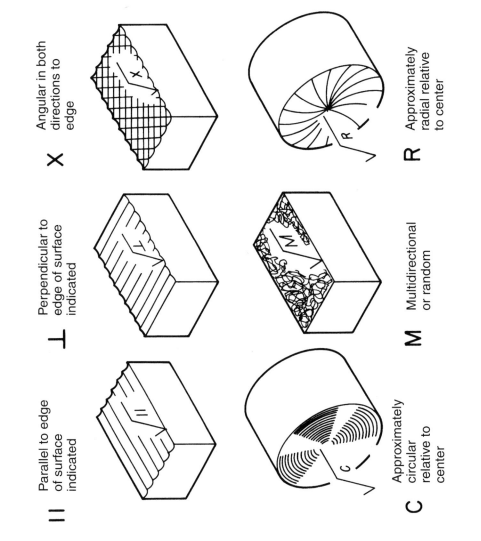

‖   Parallel to edge of surface indicated

⊥   Perpendicular to edge of surface indicated

X   Angular in both directions to edge

C   Approximately circular relative to center

M   Multidirectional or random

R   Approximately radial relative to center

Lay symbols are located beneath the horizontal bar on a surface texture symbol.

Typical Surface Finishes

Machine process	Machining finishes/microinches								
	500	250	125	63	43	16	8	4	2
Abrasive cutoff									
Automatic screw machine									
Bore									
Broach									
Counterbore									
Countersink									
Drill									
Drill (center)									
Face									
File									
Grind, cylindrical									
Grind, surface									
Hone, cylindrical									
Hone, flat									
Lap									
Mill, finish									
MIll, rough									
Ream									
Saw									
Shape									
Spotface									
Super finish									
Turn, smooth									
Turn, diamond									
Turn, rough									

The finer the finish (the lower the roughness value in microinches),
the higher the cost of obtaining it.

26-3

Metal Finishing

Name: _____ Date: _____ Score: _____

1. The symbol _____ was used at one time on drawings to   1. _____
   designate a machined surface.

2. Why was the above symbol's use discontinued? _____

   _____

   _____

3. Surface roughness is now measured in _____ and _____. What does each equal? _____

   _____

   _____

4. In addition to surface roughness, other surface conditions   4. _____
   were given values. Waviness was one such condition. It
   means:
   a. Very rough surfaces.
   b. Smoothly rounded undulations caused by tool and machine conditions.
   c. Scratches on the machined surface.
   d. All of the above.
   e. None of the above.

5. Lay is another surface finish condition. What does it mean? _____

   _____

6. A 500/12.5 surface finish is _____ than a 125/3.2 surface   6. _____
   finish.

7. While the quality of a machined surface is of paramount importance in the machining of metal,
   other finishing methods are used in the machine shop. They are employed for one or more of the
   following reasons. Explain each.

   a. Appearance: _____

   _____

   b. Protection: _____

   _____

   c. Identification: _____

   _____

   d. Cost reduction: _____

   _____

8. Regardless of the finishing method utilized, the surface to   8. _____
   be finished must be thoroughly _____ of all contaminants.

Name: _____

9. Paints, lacquers, and enamels are in the family of _____ 9. _____
   coatings.

10. List the five ways employed to apply the finishes in Question 9. _____
    _____

11. List three types of anodizing. _____
    _____
    _____

12. Describe electroplating. _____
    _____
    _____
    _____
    _____

26-4

Chapter 27

Electromachining Processes

INSTRUCTIONAL MATERIALS

Text: pages 503–510
   Test Your Knowledge Questions, page 509
Workbook: pages 143–146
Instructor's Resource: pages 341–348
   Guide for Lesson Planning
   Research and Development Ideas
   Reproducible Masters:
      27-1 The EDM Process
      27-2 The EDWC Process
      27-3 The ECM Process
      27-4 Test Your Knowledge Questions
   Color Transparency (Binder/CD only)

GUIDE FOR LESSON PLANNING

Demonstration models of electrical discharge machines can be constructed using drawings and plans appearing in various publications (trade magazines, textbooks, etc.). Research and obtain plans from these types of publications or contact manufacturers. Models should be constructed under close supervision and be inspected by a licensed electrician or electronic technician before being put into operation.

Secure examples of electrodes and work produced by electromachining processes for students/trainees to examine.

Have students read and study the chapter. Using the reproducible masters as overhead transparencies and/or handouts, review the assignment and discuss the following:

• Advantages of the electromachining processes.
• The *electrical discharge machining (EDM)* process.
• Units that make up an EDM machine.
• EDM applications.
• The *electrical discharge wire cutting (EDWC)* process.
• *Electrochemical machining (ECM)* and how it works.
• Advantages of ECM.

Technical Terms

Review the terms introduced in the chapter. New terms can be assigned as a quiz, home-work, or extra credit. The following list is also given at the beginning of the chapter.

   arc
   dielectric fluid
   electrical discharge machining (EDM)
   electrical discharge wire cutting (EDWC)

electrochemical machining (ECM)
electrode
electrolysis
electromachining
erosion
servomechanism

Review Questions

Assign *Test Your Knowledge* questions. Copy and distribute Reproducible Master 27-4 or have students use the questions on page 509 and write their answers on a separate sheet of paper.

Workbook Assignment

Assign Chapter 27 of the *Machining Fundamentals Workbook.*

Research and Development

Discuss the following topics in class or have students complete projects on their own.

1. Some of the larger tool and metalworking machinery supply houses have demonstration models of electromachining units. Contact local supply houses and borrow samples of metals worked during demonstrations along with the electrodes used.
2. Make a collection of literature on electromachining processes for the shop library.
3. Develop and produce a working model of either process. Make photos as you develop the machine and prepare a paper on the project.
4. Secure samples of work produced by EDM, EDWC, and ECM and prepare a display.

TEST YOUR KNOWLEDGE ANSWERS, Page 509

1. electrical discharge machining
2. electrical discharge wire cutting
3. electrochemical machining

4. electricity
5. chip; vaporization, microscopic particles
6. d. All of the above.
7. Dielectric fluid flushes away particles, cools the work and electrode, prevents fusion of electrode to work, and forms a nonconductive barrier.
8. servomechanism
9. mirror image
10. A metal wire is used in place of the shaped tool of EDM.
11. electroplating
12. d. All of the above.
13. rapidly
14. Refer to Section 27.3.1.

WORKBOOK ANSWERS, Pages 143–146

1. d. All of the above.
2. P
3. conduct electricity
4. d. All of the above.
5. d. All of the above.
6. a. Graphite
7. a. roughing cuts are made at low voltage and low frequency
8. faster
9. d. All of the above.
10. A small-diameter wire is used as the electrode.
11. gang-cut
12. c. there is no tool wear
13. salt, water
14. The tool does not touch the work.
15. a. one
16. Evaluate individually. Refer to Section 27.3.1.

The EDM Process

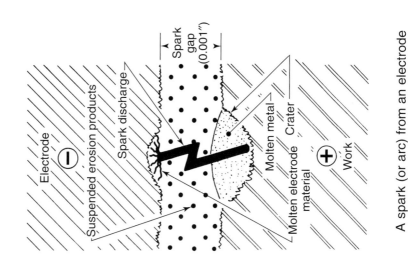

Electrode

Suspended erosion products

Spark discharge

Spark gap (0.001")

Molten metal

Crater

Molten electrode material

Work

A spark (or arc) from an electrode causes the work to erode.

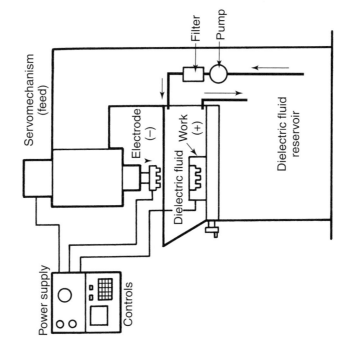

Servomechanism (feed)

Filter

Pump

Electrode (−)

Dielectric fluid

Work (+)

Dielectric fluid reservoir

Power supply

Controls

The EDWC Process

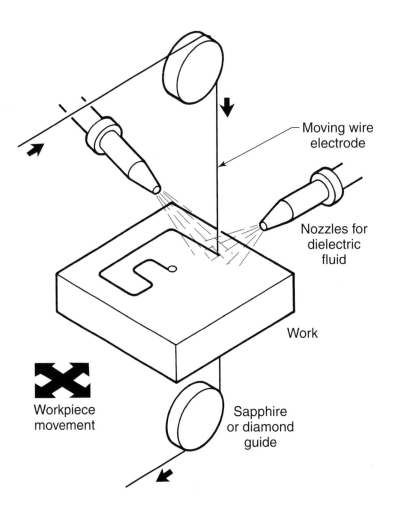

Moving wire electrode

Nozzles for dielectric fluid

Work

Workpiece movement

Sapphire or diamond guide

EDW differs from EDM in that a fine, moving wire electrode is used for cutting instead of a solid electrode. This technique is ideal for CNC operations.

27-2

The ECM Process

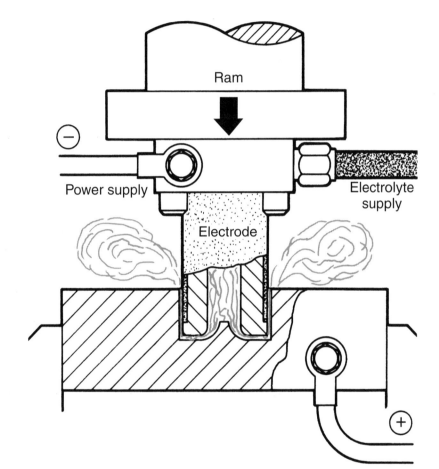

Ram

Power supply

Electrode

Electrolyte supply

27-3

Electromachining Processes

Name: _____ Date: _____ Score: _____

1. EDM stands for _____.

2. EDWC stands for _____.

3. ECM stands for _____.

4. Metals to be machined by EDM and ECM must conduct _____; otherwise, neither process can be used.

4. _____

5. Neither of the processes in Question 4 produces a _____ as metal is removed. Particles are disposed of completely by _____ or reduced to _____.

5. _____

_____

_____

6. EDM is a process that permits metals that are:
   a. Hard or tough to be machined.
   b. Difficult or impossible to machine by conventional means to be worked to close tolerances.
   c. Fragile or heat-sensitive to be machined.
   d. All of the above.
   e. None of the above.

6. _____

7. Explain the functions of the dielectric fluid used in EDM.

_____

_____

_____

8. The EDM machine _____ maintains a very thin gap of about 0.001″ (0.025 mm) between the electrode and the work.

8. _____

9. In EDM, the cavity produced in the work is an exact _____ of the electrode.

9. _____

10. How does EDWC differ from EDM? _____

_____

_____

11. ECM might well be classified as _____ in reverse.

11. _____

12. In ECM:
   a. The work is not touched by the tool.
   b. There is no friction or heat generated.
   c. There is no tool wear.
   d. All of the above.
   e. None of the above.

12. _____

27-4

(continued)

Name _____

13. In ECM, metal is removed _____ .   13. _____

14. What are five advantages that ECM offers?_____

_____

_____

_____

_____

_____

_____

Chapter 28

Nontraditional Machining Techniques

LEARNING OBJECTIVES

After studying this chapter, students will be able to:
- ○ Describe several nontraditional machining techniques.
- ○ Explain how nontraditional machining techniques differ from traditional machining processes.
- ○ Summarize how to perform several nontraditional machining techniques.
- ○ List the advantages and disadvantages of several of the nontraditional machining techniques.

INSTRUCTIONAL MATERIALS

Text: pages 511–524
    Test Your Knowledge Questions, pages 523–524
Workbook: pages 147–150
Instructor's Resource: pages 349–360
    Guide for Lesson Planning
    Research and Development Ideas
    Reproducible Masters:
        28-1 Ultrasonic Machining
        28-2 Impact Machining
        28-3 Electron Beam Welding
        28-4 Laser Beam Machining
        28-5 Test Your Knowledge Questions
    Color Transparency (Binder/CD only)

GUIDE FOR LESSON PLANNING

Have the class read and study the chapter. Using the reproducible masters as overhead transparencies and/or handouts, review the assignment, and discuss the following:
- The chemical milling process.
- Advantages and disadvantages of chemical milling.
- The chemical blanking process.
- Advantages and disadvantages of chemical blanking.
- Hydrodynamic machining (HDM).
- Ultrasonic machining.
- Ultrasonic-assist machining.
- Impact machining.
- Electron beam machining (EBM).
- Laser beam machining.

Technical Terms

Review the terms introduced in the chapter. New terms can be assigned as a quiz, homework, or extra credit. The following list is also given at the beginning of the chapter.
    chemical blanking
    chemical machining
    chemical milling
    electron beam machining
    etchant
    hydrodynamic machining
    impact (slurry) machining
    laser beam machining

ultrasonic machining
water-jet cutting

Review Questions

Assign *Test Your Knowledge* questions. Copy and distribute Reproducible Master 28-5 or have students use the questions on pages 523–524 and write their answers on a separate sheet of paper.

Workbook Assignment

Assign Chapter 28 of the *Machining Fundamentals Workbook.*

Research and Development

Discuss the following topics in class or have students complete projects on their own.

1. Prepare a file for the shop technical library on chemical milling and chemical blanking techniques. Secure literature from manufacturers of chem-milling and chem-blanking equipment and clippings from the various technical magazines.

2. Secure samples of work produced by the chemical machining techniques.

3. Develop and produce equipment that will permit you to demonstrate chemical milling. Prepare a paper on the process with photographs and submit it to one of the professional industrial education magazines.

4. Conduct a series of chemical milling experiments. Use the etchant for an equal time on different metals. Prepare a report on your experiment. List the depth of etch and what effect heat and cold have on etching rate. Develop a table showing times required to achieve equal etch depths on various metals, quality of surface finish, amount of undercut, and how it can be controlled.

5. Secure information on the use of water-jet machining.

6. Secure samples of work that have been machined using ultrasonic techniques. If the samples are small enough, mount them on a display panel. Include a sketch showing the machining technique used.

7. Gather information on other uses of ultrasonics. Prepare a bulletin board display.

8. Demonstrate how ultrasonic sound waves can be measured. Borrow a transducer and oscilloscope from the science department.

9. Construct an ultrasonic-assist. Experiment with it on the lathe.

10. Design and construct an impact machining device. Demonstrate it on various materials. Prepare an evaluation of your work.

11. Prepare a bulletin board display featuring electron beam machining. Use material from technical magazines and manufacturers' literature or brochures.

12. Prepare a research paper on electron beam machining and welding techniques. Include the history of its development and how the atomic energy, electronics, and aerospace industries use its unique characteristics.

13. Prepare a research paper on use of the laser by industry. Use illustrations from magazines. **Safety Note:** Because of the inherent dangers of using the laser, it is *not* recommended that an attempt be made to design and construct a laser capable of cutting metal.

TEST YOUR KNOWLEDGE ANSWERS, Pages 523–524

1. Chemical machining shapes metal by a selective removal of metal.

   Chemical blanking involves the total removal of metal in selected areas.

2. chem-milling, contour etching

3. In order: cleaning, masking, scribing and stripping, etching, rinsing and solvent stripping, and inspection.

4. a. Not to be etched.

5. Refer to Section 28.1.2.

6. It uses water, with abrasives added at times, under very high pressure to cut materials.

7. Slurry, impact, drilling, reaming, honing, milling, and EDM techniques use ultrasonics.

8. infrasonic

9. ultrasonic

10. special, abrasives

11. d. All of the above.

12. It is slow, the surface finish is dependent on the size of the abrasive grit used, and the deepest cut possible is 1″ (25 mm).

13. 0.001″ (0.025 mm)

14. Student answers will vary but may include the following: slicing and cutting germanium and silicon wafers; machining complex shapes in nonconductive and semiconductive materials; shaping virtually unmachinable space-age materials; improving cleaning power of chemical solvents; detecting flaws in

nondestructive testing; welding metals to nonmetals; decontaminating work that has been exposed to radioactive solutions and gases.

15. d. All of the above.
16. 0.0002″ (0.005 mm)
17. a. Thermal.
18. d. All of the above.
19. Movement of the worktable and deflection of the electron beam.
20. Laser stands for Light Amplification by Stimulated Emission of Radiation
21. Evaluate individually. Refer to Section 28.5.

WORKBOOK ANSWERS,
Pages 147–150

1. Chemicals, usually in an aqueous (with water) solution, are employed to etch away selected portions of the metal to produce an accurately contoured part.
2. masks, coating material
3. Student answers will vary, evaluate individually. Refer to Section 28.1.1.
4. Student answers will vary, evaluate individually. Refer to Section 28.1.1.
5. Chemical blanking involves complete removal of metal from certain areas by chemical action. It is a variation of chemical milling.
6. Any three of the following: tooling costs are low, no burrs are produced, new designs can be produced quickly, ultrathin metal foils can be worked, metal characteristics have no significant effect on the process.
7. water-jet
8. To shape composites of a tough fabric-like material bonded together into three-dimensional shapes called layups.
9. metals, nonmetallic
10. d. All of the above.
11. Ultrasonic-assist machining applies sound waves to the tool or metal as it is cutting or being cut. The process reduces tool forces and almost completely eliminates tool chatter.
12. c. 25,000
13. beam
14. high vacuum
15. 0.0005″ (0.0125 mm)
16. a. off longer than it is on
17. Cut geometry is controlled by movement of the worktable in the vacuum chamber and by employing the deflection coil to bend the beam of electrons to the desired cutting path.
18. intense, microns
19. 75,000°F (41 650°C)
20. b. concentrates heat in localized areas

Ultrasonic Machining

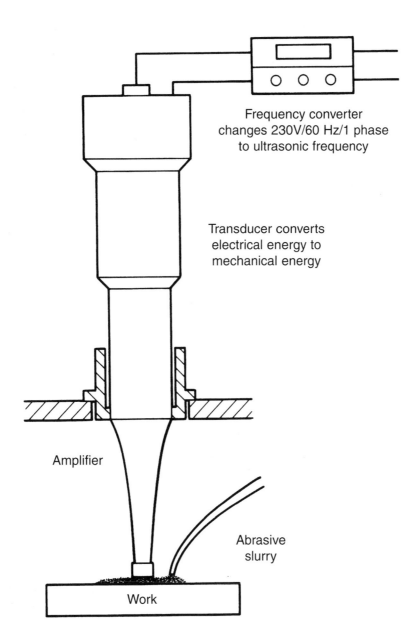

Frequency converter
changes 230V/60 Hz/1 phase
to ultrasonic frequency

Transducer converts
electrical energy to
mechanical energy

Amplifier

Abrasive
slurry

Work

28-1

Impact Machining

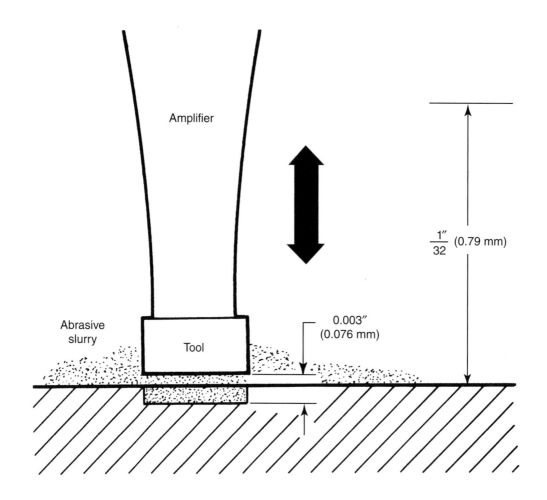

Amplifier

Abrasive
slurry

Tool

0.003″
(0.076 mm)

$\frac{1''}{32}$ (0.79 mm)

Tool motion in ultrasonic (impact) machining is slight, only 0.003″ (0.076 mm).
the 1/32″ (0.79 mm) measurement is used to indicate scale.

28-2

Electron Beam Welding

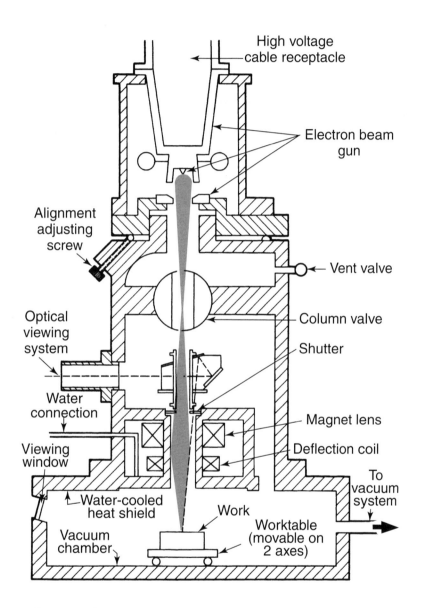

Cross-sectional view of an electron beam microcutter-welder.

28-3

Laser Beam Machining

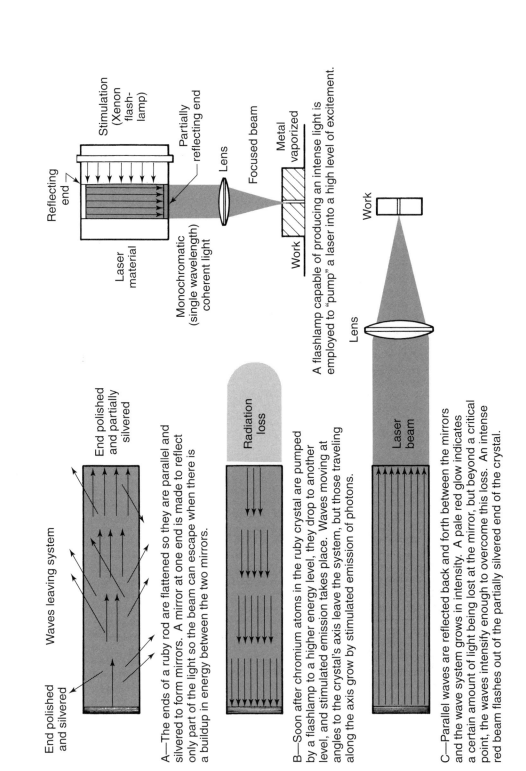

Stimulation (Xenon flash-lamp)

Reflecting end

Partially reflecting end

Laser material

Monochromatic (single wavelength) coherent light

Lens

Focused beam

Metal vaporized

Work

A flashlamp capable of producing an intense light is employed to "pump" a laser into a high level of excitement.

Lens

Work

End polished and silvered

Waves leaving system

End polished and partially silvered

A—The ends of a ruby rod are flattened so they are parallel and silvered to form mirrors. A mirror at one end is made to reflect only part of the light so the beam can escape when there is a buildup in energy between the two mirrors.

Radiation loss

B—Soon after chromium atoms in the ruby crystal are pumped by a flashlamp to a higher energy level, they drop to another level, and stimulated emission takes place. Waves moving at angles to the crystal's axis leave the system, but those traveling along the axis grow by stimulated emission of photons.

Laser beam

C—Parallel waves are reflected back and forth between the mirrors and the wave system grows in intensity. A pale red glow indicates a certain amount of light being lost at the mirror, but beyond a critical point, the waves intensify enough to overcome this loss. An intense red beam flashes out of the partially silvered end of the crystal.

Nontraditional Machining Techniques

Name: _____ Date: _____ Score: _____

1. Chemical machining falls into two categories. Briefly describe each of them. _____

_____

_____

_____

2. Chemical milling is also known as _____ or _____.   2. _____

_____

3. List the six major steps in chemical milling. _____

_____

_____

4. A mask protects the portion of a chemically milled job   4. _____
   that is:
   a. Not to be etched.
   b. To be etched.
   c. To be cleaned.
   d. All of the above.
   e. None of the above.

5. List the five major steps in chemical blanking.

_____

_____

_____

_____

_____

6. Briefly describe water-jet machining. _____

_____

_____

7. What machining processes use ultrasonics? _____

_____

8. Sound waves below 20 cycles per second are called _____ .   8. _____

9. Sound waves above 20,000 cycles per second are called   9. _____
   _____ .

10. Impact machining makes use of a _____ tool that forces   10. _____
    _____ against the work to do the cutting.

_____

28-5

(continued)

Name: _____

11. Impact machining is one of the very few commercially feasible methods for machining which types of materials?
    a. Hard.
    b. Brittle.
    c. Frangible.
    d. All of the above.
    e. None of the above.

11. _____

12. What are three disadvantages of impact machining? _____

_____

_____

13. With impact machining, tolerances of _____ can be main-tained on hole size and geometry in most materials.

13. _____

14. List five areas where the science of ultrasonics has found industrial applications.

_____

_____

_____

_____

_____

15. The development of the electron beam machine was the direct result of the special needs of what industry?
    a. Electronics.
    b. Atomic energy.
    c. Aerospace.
    d. All of the above.
    e. None of the above.

15. _____

16. Holes as small as _____ in diameter can be drilled using the electron beam technique.

16. _____

17. The electron beam machine is basically a source of what type of energy?
    a. Thermal.
    b. Sonic.
    c. Fluid.
    d. All of the above.
    e. None of the above.

17. _____

18. The electron beam technique cuts material by:
    a. Alternately heating and cooling the area to be cut.
    b. Vaporizing the material.
    c. Making use of a pulsing technique.
    d. All of the above.
    e. None of the above.

18. _____

28-5

(continued)

Name: _____

19. List two methods employed to control the shape of the cut with EBM. _____

_____

20. What does *LASER* stand for? _____

_____

21. Describe how a laser operates. _____

_____

_____

_____

Chapter 29

Other Processes

LEARNING OBJECTIVES

After studying this chapter, students will be able to:
- ○ Discuss the general machining characteristics of various plastics.
- ○ Describe the hazards associated with machining plastics.
- ○ Sharpen cutting tools to machine plastics.
- ○ Describe the five basic operations of chipless machining and their variations.
- ○ Explain how the Intraform process differs from other chipless machining techniques.
- ○ Describe how powder metallurgy parts are produced.
- ○ Relate how powder metallurgy parts can be machined.
- ○ Compare the advantages and disadvantages of various HERF techniques.
- ○ Explain how the science of cryogenics is used in industry, and list some applications.

INSTRUCTIONAL MATERIALS

Text: pages 525–546
    Test Your Knowledge Questions, pages 544–546
Workbook: pages 151–154
Instructor's Resource: pages 361–376
    Guide for Lesson Planning
    Research and Development Ideas
    Reproducible Masters:
        29-1 Chipless Machining
        29-2 Intraform® Process
        29-3 Powder Metallurgy Process
        29-4 Electrohydraulic Forming
        29-5 Pneumatic-Mechanical Forming
        29-6 Test Your Knowledge Questions
    Color Transparency (Binder/CD only)

GUIDE FOR LESSON PLANNING

This chapter presents a brief overview of several nontraditional machining processes. As each process is presented, have examples of machined products or materials on hand for class examination.

Due to the amount of material covered, it would be advisable to divide this chapter into several segments. Although it has been divided into four parts here, each classroom situation will dictate what division would work best.

Part I—Machining Plastics

Have the class read and study pages 525–531, paying particular attention to the illustrations. Review the assignment and discuss the following:

- An overview of the various plastics and their characteristics.

- Preventing heat buildup when machining plastics.
- Machining the various plastics.
- Why some plastics must be annealed.
- Sharpening cutting tools used for machining plastics.

Part II—Chipless Machining

Have the class read and study pages 531–532. Using Reproducible Masters 29-1 and 29-2, review the assignment and discuss the following:

- An overview of the various chipless machining techniques.
- Use of cold heading technique for making bolts, nuts, screws, and other fasteners.
- The Intraform® machining process.

Part III—Powder Metallurgy (P/M)

Have the class read and study pages 532–537. Using Reproducible Master 29-3, review the assignment and discuss the following:

- Powder metallurgy (P/M) applications.
- The powder metallurgy process.
- Briquetting, sintering, and forging.

Part IV—High-Energy-Rate Forming (HERF)

Have the class read and study pages 537–543. Using Reproducible Masters 29-4 and 29-5, review the assignment and discuss the following:

- An overview of the various high-energy-rate forming processes. (Electrohydraulic, magnetic, and pneumatic-mechanical forming.)
- Stand-off and contact operations.

Part V—Cryogenics

Have the class read and study pages 543–544. Review the assignment and discuss the following:

- Cryogenic applications.
- Treatment of cutting tools.

Technical Terms

Review the terms introduced in the chapter. New terms can be assigned as a quiz, homework, or extra credit. The following list is also given at the beginning of the chapter.

briquetting
chipless machining
cold heading
cryogenic
electrohydraulic forming
explosive forming
high-energy-rate forming
magnetic forming
powder metallurgy
sintering

Review Questions

Assign Test Your Knowledge questions. Copy and distribute Reproducible Master 29-6 or have students use the questions on pages 544–546 and write their answers on a separate sheet of paper.

Workbook Assignment

Assign Chapter 29 of the *Machining Fundamentals Workbook.*

Research and Development

Discuss the following topics in class or have students complete projects on their own.

Machining Plastics

1. Contact plastics manufacturers and request pamphlets on recommended machining techniques and safety precautions. Place the accumulated material in the technical library.
2. Secure samples of various plastics and demonstrate recommended machining techniques. Point out the differences between machining plastics and metal.
3. Develop a safety program to be followed when machining plastics. It can be in the form of a bulletin board display, pamphlet, or series of safety posters.
4. Review the various metalworking technical magazines and make photocopies of the many uses of plastics in the machine shop. Prepare a term paper on your findings.
5. Visit a machine shop that works plastics. Prepare a term paper on your observations. If possible, secure samples of the products produced.

High-Energy-Rate Forming (HERF)
Students should *not* experiment with explosive forming!

1. Develop a slide presentation showing step-by-step how the various HERF techniques work.

Chapter 29 Other Processes

2. Secure information on the various HERF techniques from trade journals and companies making use of HERF. Use the material to create a bulletin board display.

3. Get samples of work produced by HERF. If there are no such companies in your area, use photos from trade magazines and brochures to develop a display panel. Use a sketch of the HERF process employed to produce the particular pieces displayed or pictured.

4. Contact a company using HERF techniques. Request the loan of a film, video tape, or slides that could be used to illustrate HERF.

Chipless Machining

1. Secure samples of work produced by chipless machining. Mount them on a display panel with an illustrated explanation of the process.

2. What does the term *plasticity* mean as it relates to metal and chipless machining?

3. Secure material from companies that use the chipless machining process for the shop technical library. Prepare a display.

4. Contact a company that uses chipless machining and request samples of a product in various stages of manufacture. Prepare a display panel of the samples.

Powder Metallurgy

1. Secure a bearing and a fuel filter made using powder metallurgy technology. Examine the units under a microscope and:
   a. Make a sketch with exaggerated details that shows the structure of each example.
   b. Prepare a transparency of the sketch for use with the overhead projector. Use the projected image to explain your findings to the class.
   c. Have a microphotograph made of the grain structure of each part. Use this to illustrate a presentation on powder metallurgy.

2. Contact a firm that manufactures products using powder metallurgy and request samples of units in the various stages of the manufacturing process. Prepare a bulletin board display.

3. Secure samples of different products made by the powder metallurgy process. Prepare a display panel showing these products and how they are used. For example, fuel filters

made by the process have the ability to separate water from gasoline.

Cryogenic Applications

1. Demonstrate shrink fitting two metal parts together. Use dry ice to cool the part. **Caution:** Handle dry ice with insulated gloves and wear protective eyewear and clothing. Dry ice can cause severe burns if *not* handled with caution.

2. Read technical, scientific, and popular magazines for information on cryogenic applications. Since the aerospace and electronics industries make use of cryogenic applications, publications or companies in these fields may provide considerable amounts of material. Prepare a written or oral report on your findings.

TEST YOUR KNOWLEDGE ANSWERS, Pages 544–546

1. a. Nylon
   b. Delrin
   c. Teflon
   d. Lucite and Plexiglas
2. sharp
3. To prevent the first few threads from tearing.
4. annealing
5. chips, distorted
6. Its low coefficient of friction makes it an excellent bearing surface.
7. The dust and fumes given off by some plastics may be irritating to the skin, eyes, and respiratory system. Other plastics have fillers such as asbestos or glass fibers that can be harmful to your health. A dust collector system and filtered dust mask or respirator must be used for operator safety.
8. Laminated plastics consist of layers of reinforcing materials that have been impregnated with synthetic resins, and the layers cured with heat and pressure.
9. Drilling parallel with the laminations should be avoided.
10. cold heading, cold forming
11. There is very little scrap and production speed is increased.
12. dies
13. bolts, nuts, screws, fasteners

14. Any order: forward extrusion, backward extrusion, upsetting, trimming, piercing.
15. inside, cylindrical
16. b. Rifle barrels.
17. P/M, powder metals
18. d. All of the above.
19. In order: mixing metal powders, briquetting or forming, sintering, forging/sizing/coining.
20. d. All of the above.
21. The powder metal part before sintering. When ejected from the die it is quite brittle and fragile and must be handled carefully.
22. Because of shrinkage and distortion caused by the heating operation in the production cycle.
23. Presses the sintered pieces into precise finished dimensions, higher densities, and smoother surface finishes.
24. High-Energy-Rate Forming
25. b. In microseconds, with pressure generated by the sudden application of large amounts of energy.
26. d. All of the above.
27. spring back, HERF
28. Explosive forming makes use of the pressure wave generated by an explosion in a fluid to force the metal against the walls of the die.
29. Any of the following: cannot always form part properly on first shot; noise can be a problem; laws prohibit use of explosives in populated areas; isolated location increases transportation and handling costs; personnel must be highly skilled; high insurance rates.
30. A vacuum is necessary between the work and the die; otherwise, an air cushion would develop, preventing the metal from seating in the die and assuming its proper shape.
31. Stand-off operations and contact operations.
32. Electrohydraulic forming, electricity
33. Electromagnetic forming/magnetic pulse forming.
34. shrink, expand
35. high-pressure gas, punch, die
36. It means to make icy cold.
37. –300°F (–184°C), –460°F (–293°C)
38. One part of the assembly is made slightly oversize and immersed in liquid nitrogen. The diameter is reduced (shrunk) by the extreme temperature drop until it fits easily into its mating part. As it returns to room temperature, the cooled part expands and is locked in place.
39. The parts do not become distorted as they would if they were mechanically pressed together or heated (expanded).
40. To prevent damage from thermal shock.

WORKBOOK ANSWERS, Pages 151–154

1. Dust and fumes given off by some plastics may be irritating to the skin, eyes, and respiratory system. Other plastics have fillers such as asbestos and glass fibers which are harmful to your health.
2. Any of the following: high tensile strength, impact resistant, flexible strength, resistance to abrasion, not affected by most chemicals, greases, and solvents.
3. supported
4. soft brass
5. Tools must be kept sharp to prevent the plastic from melting or becoming gummy. Sharp tools also assure a good quality surface finish.
6. annealing
7. It has excellent dimensional stability, high strength, and rigidity. It has low friction, requires minimal use of lubricants, and is very quiet in operation. It is replacing brass and zinc for many applications in the automotive and plumbing industries and is used for parts in business machines.
8. a. the temperature at which it will be used
9. They prevent the heat from dissipating.
10. To deter thermal expansion.
11. Chipless machining forms wire or rod into the desired shape using a series of dies.
12. Cost saving on some jobs, scrap is reduced, and increased production speed.
13. spark plug
14. Sintering is another name for powder metallurgy, a technique used to shape parts from metal powders. It is also the process of transforming the briquette into a strong unit.
15. They are brittle and very fragile.
16. c. the metal tries to return to its original shape
17. Stand-off operations: The charge is located some distance above the work. Its energy is transmitted through a fluid medium, such as water.

Contact operations: The charge is touching the work and the explosive energy acts directly on the metal.

18. An insulated coil is wrapped around or placed within the work. As very high momentary current is passed through the coil, an immense magnetic field is created causing the work to collapse, compress, shrink, or expand depending upon the design of the coil.

19. High-pressure gas is used to accelerate a punch into a die. The forces developed are many times more powerful than those used in conventional forging and are sufficient to shape hard-to-work materials. The metal blank is heated prior to the forming operation and the machine requires less space than the conventional forging press.

20. b. the use of liquid nitrogen as a gas

Chipless Machining

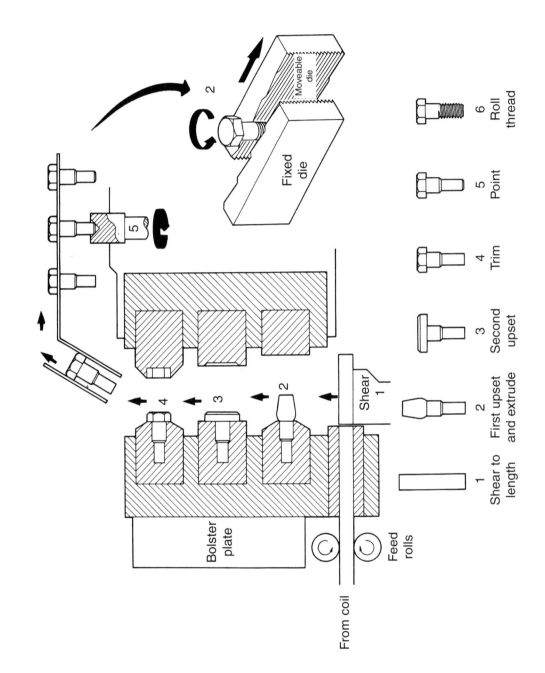

Intraform® Process

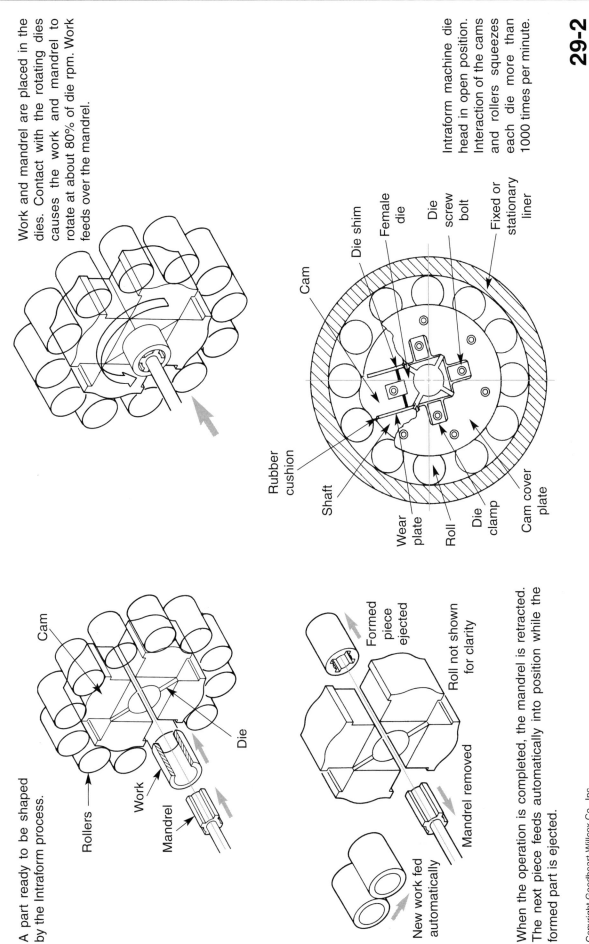

A part ready to be shaped by the Intraform process.

Cam

Rollers

Work

Mandrel

Die

Work and mandrel are placed in the dies. Contact with the rotating dies causes the work and mandrel to rotate at about 80% of die rpm. Work feeds over the mandrel.

Cam

Rubber cushion

Shaft

Wear plate

Roll

Die clamp

Cam cover plate

Die shim

Female die

Die screw bolt

Fixed or stationary liner

Intraform machine die head in open position. Interaction of the cams and rollers squeezes each die more than 1000 times per minute.

Formed piece ejected

Roll not shown for clarity

Mandrel removed

New work fed automatically

When the operation is completed, the mandrel is retracted. The next piece feeds automatically into position while the formed part is ejected.

Powder Metallurgy Process

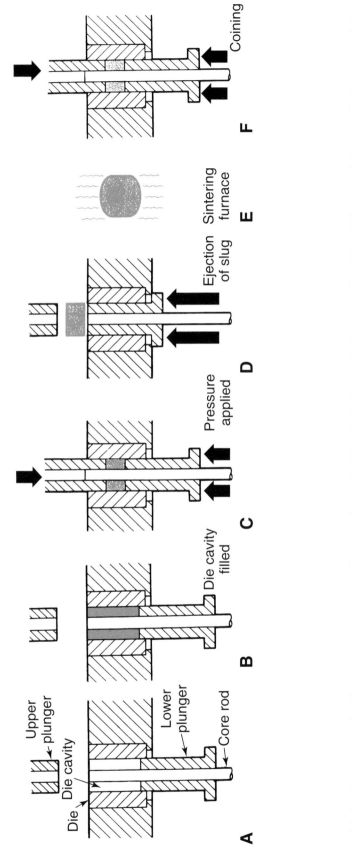

29-3

Steps in fabricating a part. A—Note cross section of die and die cavity. Depth of the cavity is determined by thickness of the required part, and the amount of pressure that will be applied. B—Die cavity is filled with proper metal powder mixture. C—Pressure as high as 50 tons per square inch is applied. D—Briquette or "slug" is pushed from die cavity. E—Pieces are then passed through a sintering furnace to convert them into a strong, useful product. F—Some pieces can be used as they come from the furnace. Others may require a coining or sizing operation to bring them to exact size and to improve their surface finish.

Electrohydraulic Forming

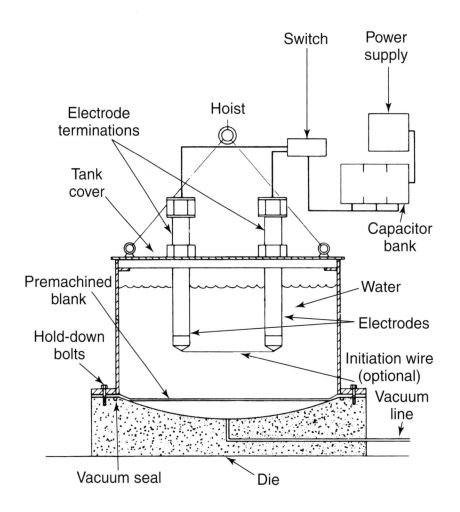

Diagram shows setup for electrohydraulic forming, which uses electrical energy as a source of power for HERF operations.

29-4

Pneumatic-Mechanical Forming

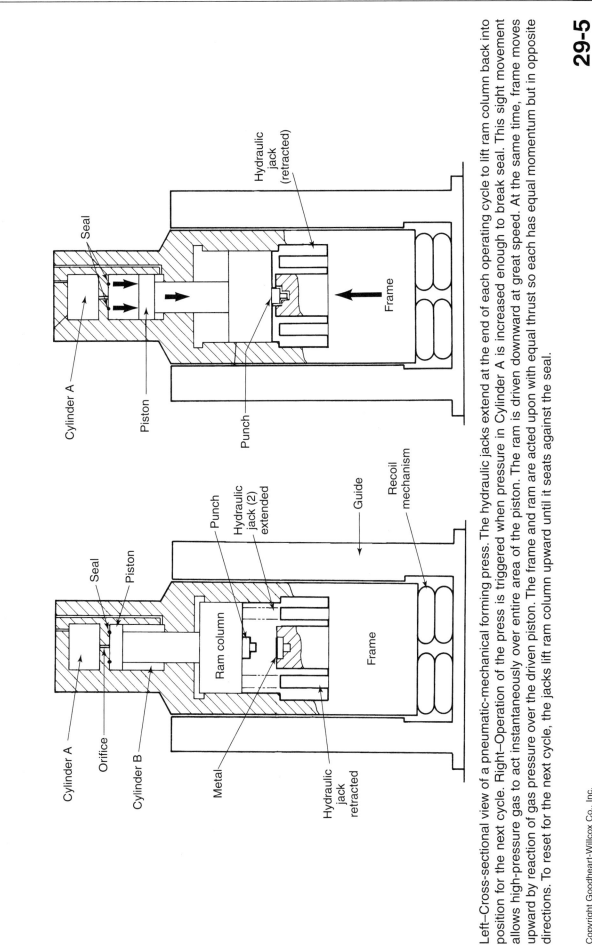

29-5

Left—Cross-sectional view of a pneumatic-mechanical forming press. The hydraulic jacks extend at the end of each operating cycle to lift ram column back into position for the next cycle. Right—Operation of the press is triggered when pressure in Cylinder A is increased enough to break seal. This sight movement allows high-pressure gas to act instantaneously over entire area of the piston. The ram is driven downward at great speed. At the same time, frame moves upward by reaction of gas pressure over the driven piston. The frame and ram are acted upon with equal thrust so each has equal momentum but in opposite directions. To reset for the next cycle, the jacks lift ram column upward until it seats against the seal.

Other Processes

Name: _____ Date: _____ Score: _____

1. Give a common trade name for each of the following types of plastics.
   a. Polyamide resins: _____

   b. Acetal resins: _____

   c. Fluorocarbon resins: _____

   d. Acrylic resins: _____

2. If plastics are to be machined with any degree of accuracy, the cutting tools must be _____.

   2. _____

3. When hand-threading plastics, why is it recommended that the hole or rod be chamfered?

   _____

4. Like metal, many plastics require _____ to ensure against dimensional changes.

   4. _____

5. When turning many plastics on a lathe, care must be taken to prevent the _____ from accumulating around the work. If this is not done, heat will build up and cause the plastic to become _____.

   5. _____

   _____

6. What is unique about Teflon®? _____

   _____

7. Machining plastics can create health problems for the machinist if precautions are not taken. What are these problems and how can they best be handled? _____

   _____

   _____

8. What are laminated plastics? _____

   _____

   _____

9. When drilling laminated plastics, what should be avoided? _____

   _____

10. Chipless machining is also known as _____ or _____.

    10. _____

    _____

11. How does chipless machining make substantial savings possible? _____

    _____

12. In chipless machining a series of _____ replaces the usual cutting tools of the lathe, drill press, and milling machine.

    12. _____

Name: _____

13. Chipless machining is still the most economical way to make _____, _____, _____, and other types of _____.

13. _____

_____

_____

_____

14. List the five basic operations performed by machines making use of the chipless machining process._____

_____

15. *Intraform®* is a chipless machining technique that can form profiles on the _____ of _____ pieces.

15. _____

_____

16. The Intraform® technique has proven to be a practical way to produce:
    a. Socket wrenches.
    b. Rifle barrels.
    c. Automotive starter clutch housings.
    d. All of the above.
    e. None of the above.

16. _____

17. Powder metallurgy, abbreviated _____, is the technique of shaping parts from _____.

17. _____

_____

18. The powder metallurgy process is used to make:
    a. Self-lubricating bearings.
    b. Precision machine parts.
    c. Permanent metal filters.
    d. All of the above.
    e. None of the above.

18. _____

19. List the steps in making a part by the powder metallurgy technique. _____

_____

_____

20. Parts made from metal powder can be:
    a. Drilled.
    b. Heat treated.
    c. Turned on a lathe.
    d. All of the above.
    e. None of the above.

20. _____

21. What is a briquette or "green compact?" _____

_____

_____

29-6

(continued)

Name: _____

22. Why is it often necessary to size, coin, or forge parts made from metal powders after they have been sintered? _____

_____

23. What do the above operations do to the sintered piece? _____

_____

24. The abbreviation HERF means _____

_____.

25. In HERF, metal is shaped: 25. _____
    a. By the slow application of great pressure.
    b. In microseconds, with pressure generated by the
       sudden application of large amounts of energy.
    c. By conventional forging methods.
    d. All of the above.
    e. None of the above.

26. The pressures needed in HERF are generated by: 26. _____
    a. Detonating explosives.
    b. Releasing compressed gases.
    c. Electromagnetic energy.
    d. All of the above.
    e. None of the above.

27. Many metals tend to _____ to their original shape after   27. _____
    being formed by conventional means. This problem is
    greatly reduced or entirely eliminated when _____ is   _____
    used to shape the metal.

28. What is explosive forming? _____

_____

_____

29. What are some of the disadvantages of explosive forming? _____

_____

_____

_____

30. Why must a vacuum be pulled in the die when explosive forming? _____

_____

_____

31. Depending upon the placement of the explosive, most explosive forming operations fall into two categories. List them. _____

_____

Name: _____

32. _____ is a variation of explosive forming. However, _____ is used in place of the explosive charge to generate the required energy.

32. _____

_____

33. What HERF technique employs a very high electric current passing through an induction coil shaped to produce the required configuration in the work?

33. _____

34. The technique in Question 33 can be used to _____ or _____ on the work to produce the desired shape, depending upon placement of the coil.

34. _____

_____

35. In pneumatic-mechanical forming, _____ is used to accelerate the _____ into the _____.

35. _____

_____

_____

36. What does the term "cryogenic" mean? _____

_____

37. The science of cryogenics deals with temperature starting at _____ (_____) and goes down to temperatures near _____ (_____).

37. _____

_____

_____

_____

38. What does shrink fitting mean? _____

_____

_____

_____

39. Why is it better to use the super-low temperatures of cryogenics, rather than heat, to shrink-fit parts together? _____

_____

_____

40. Why must the cooling of treated cutting tools be done at a slow rate? _____

_____

Chapter 30

Occupations in Machining Technology

LEARNING OBJECTIVES

After studying this chapter, students will be able to:
- List the requirements for the various machining technology occupations.
- Explain where to obtain information on occupations in machining technology.
- State what industry expects of an employee.
- Describe what an employee should expect from industry.
- Summarize the information given on a resume.

INSTRUCTIONAL MATERIALS

Text: pages 547–556
   Test Your Knowledge Questions, page 556
Workbook: pages 155–160
Instructor's Resource: pages 377–384
   Reproducible Masters:
      30-1 Job Application
      30-2 Test Your Knowledge Questions
   Color Transparency (Binder/CD only)

GUIDE FOR LESSON PLANNING

Have the class read and study the chapter. Review the assignment and discuss the following:

- The four general categories of jobs in machining technology. The advantages and disadvantages, education and/or training required, type of work, and salary scale for each category.
- Where to get information on machining technology.
- How to prepare a resume and the reasons for doing so.
- How to accurately fill out a job application.
- How to prepare for a job interview.
- Factors that can lead to job termination.

The importance of accurately completing a job application *cannot* be overemphasized. A resume prepared beforehand will prove very helpful.

Have a selection of job applications for the class to examine so they can see what information is usually requested.

Students/trainees should be aware that obtaining employment in the machining technology fields will require their full attention. After locating a position, encourage students to learn as much as possible about the company before sending their resume.

Filling out a job application is the next step in securing employment. After that, there is usually a personal interview. Help students prepare for the interview by discussing proper dress, attitude, and the types of things that can cause a poor impression. Use Reproducible Master 30-1 as a practice job application. (This is a duplicate of the application in the workbook.)

Use the chalkboard and make two lists. One list should show what makes a good interview. The other list should include what can cause a poor interview. Have students/trainees indicate what they think would fall into both categories.

Mock interviews are often used to provide experience and build student/trainee confidence. Use an outside source as the interviewer.

Invite a person from the local state employment service to speak to the class on job opportunities on the local and state levels. Request that they cover education/training requirements, starting salaries, and how to locate these types of jobs.

Technical Terms

Review the terms introduced in the chapter. New terms can be assigned as a quiz, homework, or extra credit. The following list is also given at the beginning of the chapter.

> *all-around machinist*
> *apprentice programs*
> *career*
> *engineering*
> *job shops*
> *part programmer*
> *resume*
> *semiskilled workers*
> *skilled workers*
> *technician*

Review Questions

Assign *Test Your Knowledge* questions. Copy and distribute Reproducible Master 30-2 or have students use the questions on page 556 and write their answers on a separate sheet of paper.

Workbook Assignment

Assign Chapter 30 of the *Machining Fundamentals Workbook*. Have students complete the job application on pages 159–160.

Research and Development

Discuss the following topics in class or have students complete projects on their own.

1. Invite a speaker from the local government employment office (it may be listed under different names in various parts of the country) to discuss local and national employment opportunities in the metalworking industries.
2. Prepare a chart that shows the hourly salaries of skilled and technical machine shop employees at local industries.
3. Ask someone who has graduated from your program and who has completed an apprentice program to describe their experiences.
4. Prepare a bulletin board display around a local apprentice program. List the training requirements on a year-to-year basis.
5. Summarize the information on machine shops and related occupations described in the occupational outlook handbook (a government publication) and make it available to the class.
6. Contact the International Association of Machinists for information on apprentice programs.
7. Study the help wanted columns in your local papers. Collect the advertisements in the metalworking trades. What types of jobs are available? How do the various employers advertise? Do they offer fringe benefits? Are they equal opportunity employers? Do they pay a premium for second- and third-shift workers? Make this study every three months and prepare a chart that will show whether the demand increases or decreases during the various seasons.
8. Secure a selection of job applications. Study them carefully and discuss with the class the information they request. What can be done to make applying for a job easier and simpler?

TEST YOUR KNOWLEDGE ANSWERS, Page 556

1. Semiskilled, skilled, technical, and professional.
2. Semiskilled
3. skilled
4. Armed Forces, vocational/technical programs in high schools and community colleges.
5. They are expected to plan and carry out all of the operations needed to machine a job.
6. Interprets drawings and, with precision measuring tools, locates and marks off where metal must be removed by machining from castings, forgings, and metal stock.
7. Formal training in computer technology as it relates to machine tool operation; experience in reading and interpreting drawings; thorough knowledge of machining technology and procedures; a working knowledge of cutting speeds and feeds for various types of machine tools and materials; and training in mathematics.
8. The technician assists the engineer by testing various experimental devices and machines,

compiling statistics, making cost estimates, and preparing technical reports. Many inspection and quality control programs are managed by technicians. Technicians also repair and maintain computer controlled machine tools and robots.

9. The job usually requires at least two years of college, with a program of study centered on math, science, English, computer science, quality control, manufacturing, and production processes.

10. Student answers will vary but may include any three of the following: state employment service, school's career center, teachers of industrial/technical departments, Armed Forces, and the Department of Labor.

11. Student answers will vary, evaluate individually.

12. A summary of your educational and employment background.

13. It will assure uniform information with little chance for confusing responses.

14. Student answers will vary, evaluate individually.

15. Refer to Section 30.2.2.

16. Student answers will vary but may include three of the following:

Alcohol and/or illegal drug abuse on the job.

Inability or refusal to perform the work required.

Being habitually tardy or missing work repeatedly without adequate reasons.

Inability to work with supervisors or peers.

Fighting with or making threats to fellow workers or supervisors.

Inability to work as a team member.

WORKBOOK ANSWERS,
Pages 155–160

1. do not require a high degree of skill or training

2. additional study, training

3. apprentice, four

4. d. All of the above.

5. Specializes in producing tools, dies, and fixtures that are necessary for modern mass-production techniques.

6. Toolmaker who specializes in making the punches and dies needed to stamp out such parts as auto body panels and electrical components. Also produces dies for making extrusions and die castings.

7. Person who locates and positions tooling and work-holding devices on a machine tool for use by a machine tool operator. This worker may also show the machine tool operator how to do the job, and often checks the accuracy of the machined part.

8. Usually a skilled machinist who has been promoted to a position of greater responsibility. This person will direct other workers in the shop and is responsible for meeting production deadlines and keeping work quality high. In many shops, the manager may also be responsible training and other tasks

9. Inputs data into computer-controlled (CNC) machine tool for machining a product by determining the sequences, tools, and motions the machine must carry out to machine the part.

10. d. All of the above.

11. Student answers will vary, evaluate individually.

12. Student answers will vary, evaluate individually.

13. Refer to Section 30.2.3.

14. Additional activities A–C, involve instructor's permission and/or group participation. Review activities and assign as applicable or as time permits.

15. The completed job application should be evaluated individually.

Job Application

Please print all information. You must fully and accurately complete the application.

PERSONAL INFORMATION

Date   Social Security Number

Name
Last   First   Middle

Present Address
Street   City   State

Permanent Address
Street   City   State

Phone No.   Alternate No.

If related to anyone in our employ, state name and department   Referred by

EMPLOYMENT DESIRED

Position   Date you can start   Salary desired

Are you employed now?   If so may we inquire of your present employer?

Ever applied to this company before?   Where   When

EDUCATION

	Name and Location of School	Years Completed	Subjects Studied	Degree Earned
Grammar School				
High School				
College/University				
Trade, Business or Correspondence School				

Subject of special study or research work

What foreign languages do you speak fluently?   Read fluently?   Write fluently?

U.S. Military service   Rank   Present membership in National Guard or Reserves

Name: _____

Activities other than religious (Exclude organizations the name or character of which indicates the race, creed, color or national origin of its members.)

FORMER EMPLOYERS List your employers for the past five years, beginning with the last (or present) employer first.

Date

Month and Year	Name and Address of Employer	Salary	Position	Reason for Leaving
From _____				
To				
From _____				
To				
From _____				
To				

REFERENCES Give below the names of two persons not related to you whom you have known for at least one year

	Name	Address	Job Title	Years Acquainted
1				
2				

PHYSICAL RECORD

Have you any disabilities that might affect your ability to perform this job? _____

In case of
emergency notify _____

         Name   Address   Phone No.

We are an equal opportunity employer. We are dedicated to a policy of non-discrimination in employment on any basis including race, creed, age, sex, religion, national origin, height, weight, marital status, or disability.

I understand that to accept employment, I must be lawfully authorized to work in the United States, and I must present documents to prove my eligibility.

I understand that the company may thoroughly investigate my work and personal history and verify all data given on this application, on related papers, and in interviews. I authorize all individuals, schools, and firms named therein, except my current employer if so noted, to provide any information requested about me, and I release them from all liability for damage in providing this information.

The information on this application and any made in conjunction with this application is correct and true to the best of my knowledge. I understand that any false or misleading statement made by me in connection with this application or the failure to disclose any material will be grounds for immediate dismissal.

In consideration of my employment, I agree to conform to the rules and regulations of this company, and my employment and compensation can be terminated, with or without notice, at any time, at the option of either the company or myself. I understand that no manager or representative of the company, other than the president/owner of the company, has any authority to enter into any agreement for employment for any specified period of time, or to make any agreement contrary to the foregoing.

I authorize investigation of all statements contained in this application. I understand that misrepresentation or omission of facts called for is cause for dismissal.

Date _____ Signature _____

30-1

Occupations in Machining Technology

Name: _____ Date: _____ Score: _____

1. List the four general categories into which metalworking occupations fall. _____
   _____

2. _____ workers are those who perform operations that do   2. _____
   not require a high degree of skill or training.

3. The _____ worker usually starts his or her carrear as an   3. _____
   apprentice.

4. Since the number of apprentice programs is on the decline, where can this training now be
   obtained? _____
   _____

5. Describe what an all-around machinist is expected to do. _____
   _____
   _____

6. What does a layout specialist do? _____
   _____
   _____
   _____

7. To perform his or her job properly, a part programmer should have the following background:
   (List five items.) _____
   _____
   _____
   _____
   _____

8. What are some of the duties of a technician? _____
   _____
   _____
   _____
   _____

9. List the areas of study usually included in a technician's educational program. _____
   _____
   _____

Name: _____

10. List three sources of information on metalworking occupations. _____

_____

_____

11. What does industry expect from you when you are on the job? _____

_____

_____

_____

_____

12. What is a job resume? _____

_____

_____

_____

13. Why should a resume be prepared in advance? _____

_____

_____

_____

14. Explain why you think references are important. _____

_____

_____

_____

15. List five traits an employer wants in a prospective employee. _____

_____

_____

_____

_____

_____

16. What are three factors that can lead to job termination? _____

_____

_____

_____

Color Transparencies (Binder/CD only)

These color transparencies are designed to help your students understand the principles and techniques of machining technology. They can be used to introduce fundamentals of operation, generate discussion on proper techniques, and facilitate and improve your presentation to the class.

Chapter 1

CT1-1. The lathe operates on the principle of the work being rotated against the edge of a cutting tool.

CT1-2. A drill press operates by rotating a cutting tool (drill) against the material with sufficient pressure to cause the tool to penetrate the material.

CT1-3. Grinding is a cutting operation, like turning, drilling, milling, or sawing. However, instead of the one, two, or multiple-edge cutting tools used in other applications, grinding employs an abrasive tool composed of thousands of cutting edges.

CT1-4. Band machining makes use of a continuous saw blade, with each tooth functioning as a precision cutting tool.

CT1-5. Milling removes material by rotating a multitoothed cutter into the work. A—With peripheral milling, the surface being machined is parallel to periphery of the cutter. B—End mills have cutting edges on the circumference and the end.

CT1-6. A broach is a multitoothed cutting tool that moves against the work. Each tooth removes only a small portion of the material being machined. The cutting operation may be on a vertical or horizontal plane.

Chapter 2

CT2-1. Chart illustrates various fire extinguisher types and fire classifications. In the small engine shop, always use an extinguisher designed for use on electrical and chemical fires.

Chapter 3

CT3-1. Many types of lines, symbols, and figures are used to give a drawing exact meaning.

CT3-2. Standard ANSI symbols are changed periodically. You must be familiar with both the old and new symbols because either may be used on the drawings. Compare these examples.

CT3-3. A great deal of information is contained in the drawing's title block. The components highlighted here are standard on most drawings.

Chapter 4

CT4-1. These are the US conventional fractional and metric graduations found on rules. Measurements are taken by counting the number of graduations.

CT4-2. How to read a Vernier micrometer caliper. Add the total reading in thousandths, then observe which of the lines on the Vernier scale coincides with a line on the thimble. In this case, it is the second line, so 0.00002 is added to the reading.

CT4-3. To read a metric micrometer, add the total reading in millimeters visible on the sleeve to the reading of hundredths of a millimeter, indicated by the graduation on the thimble. Note that the thimble reading coincides with the longitudinal line on the micrometer sleeve.

CT4-4. Reading a metric-based Vernier micrometer caliper. To the regular reading in hundredths of a millimeter (0.01), add the reading from the Vernier scale that coincides with a line on the thimble. Each line on the Vernier scale is equal to two thousandths of a millimeter (00.002 mm).

CT4-5. Reading a 25- and 50-division Vernier scale.

CT4-6. A—How to read a 25-division metric-based Vernier scale. Readings on the scale are obtained in units of two hundredths of a millimeter (0.02 mm). B—How to read a 50-division metric-based Vernier scale. Each division equals two hundredths of a millimeter (0.02 mm).

Chapter 5

CT5-1. Compare the part drawing with steps involved in laying out the job.

Chapter 6

CT6-1. How thread size is noted and what each term means.

CT6-2. Nomenclature of a fastener thread.

Chapter 7

CT7-1. Cap screws are manufactured in various types. A—Flat head. B—Hexagonal head. C—Socket head. D—Fillister head. E—Button or round head.

CT7-2. Identification marks (inch size) and class numbers (metric size) are used to indicate the relative strength of hex head cap screws. As identification marks increase in number, or class numbers become larger, increasing strength is indicated.

CT7-3. Setscrew head and point designs. A—Socket head. B—Slotted head. C, D, and E—Fluted head. F—Square head. G—Flat point. H—Oval point. I—Cone point. J—Half dog point. K—Full dog point. L—Cut point.

Chapter 8

CT8-1. A simple drill template. Identification numbers on jigs and fixtures allow these devices to be located easily when stored away between uses.

Chapter 9

CT9-1. For maximum results, coolant should flood the area being machined and cutting tool to provide the most efficient removal of the heat generated. (Kesel/JRM International, Inc.)

Chapter 10

CT10-1. A—Gun drill. The tip is shown in larger scale. The light-colored portion is tungsten carbide. The larger area does the cutting; the smaller sections act as wear surfaces. B—A taper shank twist drill with holes to direct coolant to the cutting edges. C—Three- and four-flute core drills. D—Step drill. E—Combination drill and reamer. F—Microdrills are smaller than the #80 drill (0.0135″ diameter). G—A half-round drill.

Chapter 11

CT11-1. Principles of how typical grinding machines work.

Chapter 12

CT12-1. The three principle types of cutoff saws.

CT12-2. Recommended ways to hold sharp-cornered work for cutting. A carefully planned setup will ensure that at least three teeth will be cutting, greatly extending blade life.

Chapter 13

CT13-1. Lathe measurements. A—Length of bed. B—Distance between centers. C—Diameter of work that can be turned over the ways. D— Diameter of work that can be turned over the cross-slide.

CT13-2. Cutter bit nomenclature.

Chapter 14

CT14-1. Common thread forms. A—Unified thread form, interchangeable with American National Thread. B—Sharp "V" thread form. C—Acme thread form. D—Square thread form. Note: In formulas above, N = Number of threads per inch; P = Pitch; d = depth of thread.

CT14-2. The difference between lead and pitch. A—Single thread screw, the pitch and lead are equal. B—Double thread screw, the lead is twice the pitch. C—Triple thread screw, the lead is three times the pitch.

Chapter 15

CT15-1. Centering drill. A—The drill will cut exactly on center if the hole is started with a center drill. B—Holes larger than 1/2″ (12.5 mm) in diameter require drilling of a pilot hole. C—There must be enough clearance between the back of the work and the chuck face to permit the drill to break through the work without damaging the chuck.

Chapter 16

CT16-1. This drawing shows a greatly shortened section of an internal broaching tool and a cross-section of the splines it cuts in a part. The pilot guides the cutter into a cut or hole previously made in the work. Each tooth of the broach increases slightly in size until the specified size is attained.

Chapter 17

CT17-1. A—Table movements of plain-type horizontal milling machine. B—Table movements of the universal type milling machine.

Chapter 18

CT18-1. Four of the most common methods employed to mount end mills in a vertical milling machine. A—Adapter sleeve with taper shank cutter. B—B&S taper mounted directly in the spindle. C—Spring collet with straight shank cutter. D—Adapter with setscrew on straight shank cutter.

CT18-2. Sequence for squaring work on a milling machine.

CT18-3. Nomenclature of bevel gears. The smaller gear is called the pinion.

Chapter 19

CT19-1. Three variations of the planer-type surface grinder.

CT19-2. Standard system for marking grinding wheels.

CT19-3. Standard grinding wheel shapes.

CT19-4. Cutting edges of a machine reamer.

CT19-5. A few of many abrasive belt grinding techniques.

Chapter 20

CT20-1. Blade set. A—The term "blade set" refers to side angle of the teeth. B—The different types of blade set. C—Saw blade terminology.

CT20-2. Blade guides. A—Blade guide inserts are used for light sawing. B—Roller guides are recommended for continuous high-speed cutting.

Chapter 21

CT21-1. Axes of machine tool movements. A—Vertical milling machine. B—Lathe. C—Horizontal milling machine. Spindle motion is assigned Z axis. Note how Z axis differs between machines with vertical spindle and machines with horizontal spindle.

Chapter 22

CT22-1. Basic geometric configurations of robots. All provide three articulations (specific arm movements).

Chapter 23

CT23-1. A—One type of laser inspection device. With it, tool wear can be continuously monitored. Note how laser beams can detect work that is oversize or undersize. B—Operation of the eddy current flaw detection system. Photoelectric cells turn off alarm system when the end of a test piece passes into the test coil. Any flaw would cause small current changes in the test coil.

Chapter 24

CT24-1. A spark test is sometimes employed to determine the grade of steel. Touch steel to the grinding wheel lightly, and observe the color and form of the resulting sparks.

Chapter 25

CT25-1. Before metal become incandescent (glows red), steel will pass through the colors listed on this table. Colors are also useful for tempering steel if no pyrometer is fitted to the furnace. (Bethlehem Steel Co.)

CT25-2. This diagram shows how 1/16″ steel ball penetrator is used to make a Rockwell B hardness reading. Size of the ball has been greatly exaggerated for clarity.

Chapter 26

CT26-1. A—How surface waviness is measured. Note the difference in magnitude between waviness and roughness. B—On drawings, symbols and numbers show roughness, waviness, and lay. They specify finishes required on a surface.

Chapter 27

CT27-1. A—How electrical discharge wire cutting (EDWC) works. It differs from EDM in that a fine, moving wire electrode is used for cutting instead of a solid electrode. B—This technique is ideal for CNC operations. How electrochemical machining (ECM) works.

Chapter 28

CT28-1. This aircraft part was chem-milled after it was formed. Pieces even more severely formed than this part can be chem-milled economically. (Northrop-Grumman Corp.)

Chapter 29

CT29-1. Arrows and numbers indicate sequence involved in producing bolts by chipless machining. Trace part flow through the sequence of operations.

Chapter 30

CT30-1. A well-written resume will help make a good impression on employers.

NOTES

NOTES

NOTES